Meteorology, Climatology and Disaster Management
Theory Manual

THE AUTHORS

Dr. M. Divya is an Assistant Professor of the Department of Aquatic Environment Management at the College of Fisheries Science, St. Devasahayam Institute of Fisheries Science and Technology, located in Midalam, Kanniyakumari District, Tamil Nadu. With four years of research experience under UGC and DBT funded projects and two years of teaching experience, she has contributed adequately to her field. She has published 15 research papers in reputable national and international journals. She conducted a few capacity-building programs funded by NFDB, Hyderabad, and has actively participated and presented in numerous national and international conferences, symposiums, workshops, and seminars. Her expertise lies in the fields of Planktonology and Bioremediation technology.

Prof. S. Felix is a distinguished figure in the field of fisheries education and technology consulting, with a career spanning over four decades. Currently serving as the Dean of the College of Fisheries Science at St. Devasahayam Institute of Fisheries Science and Technology in Midalam, Kanniyakumari district, Tamil Nadu and also President of AQUAFIN (Aquaculture Technology & Research Foundation India). He has held several prestigious positions, including Vice Chancellor of Tamil Nadu Fisheries University (TNJFU) and President of the World Aquaculture Society (WAS)-APC. Prof. Felix is recognized for his exemplary leadership and contributions to the fisheries and aquaculture sector, demonstrated through his involvement in various committees and organizations. He is a prolific author, having published numerous research papers, books, and manuals, and has played a pivotal role in introducing new fisheries science courses and establishing fisheries colleges and research centers in the state in the state.

Meteorology, Climatology and Disaster Management

Theory Manual

— Authors —

M. Divya

S. Felix

College of Fisheries Science
St. Devasahayam Institute of
Fisheries Science & Technology,
Midalam, Kanniyakumari Dist.629193 (Tamil Nadu)

2025

Daya Publishing House®

A Division of

Astral International Pvt. Ltd.

New Delhi – 110 002

ISBN: 978-93-5919-787-6

Published by : **Daya Publishing House®**
A Division of
Astral International Pvt. Ltd.
– ISO 9001:2015 Certified Company –
4736/23, Ansari Road, Darya Ganj
New Delhi-110 002
Ph. 011-43549197, 23278134
E-mail: info@astralint.com
Website: www.astralint.com

Printed at : **Replika Press Pvt. Ltd**

Preface

Welcome to this Meteorology, Climatology and Disaster Management: Theory Manual. As the forces of nature continue to shape our world, it is imperative that we equip ourselves with the knowledge and tools needed to navigate the challenges and opportunities that lie ahead. Its purpose is to provide readers with a comprehensive understanding of the interconnectedness of meteorology, climatology, and disaster management, and to offer practical guidance for effectively addressing environmental challenges. Each chapter has been carefully crafted to cover a wide range of topics, from the basic principles of atmospheric science to advanced techniques in disaster preparedness and response. Whether you are a student looking to deepen your understanding of these disciplines, a professional seeking to enhance your skills, or a concerned citizen eager to make a difference in your community, this manual is intended to meet your needs. We would like to express our sincere thanks to publishers who have generously accepted for publication. It is our hope that this manual will serve as a valuable resource and inspire readers to take action towards building a more resilient and sustainable future.

M. Divya
S. Felix

Contents

Meteorology, Climatology and Disaster Management

- Meteorology is the science that deals with the study of the atmosphere.
- Geography studies earth as the home of man. It attempts to find out the relationship between man and his environment.
- This course is important to understand meteorological and geographical aspects of the earth that influence on all most all activities of the human being. This course imparts overall knowledge about meteorology and geography which is applicable in fisheries activities. As fisheries is concerned, knowledge of meteorology and geography is essential for fish/shrimp farming activities as well as capture fisheries. This course imparts information on weather and climate, clouds, winds, longitude, latitude, thunderstorms, cyclones, south-west and northeast monsoon which are helpful in fisheries activities.
- This course gives information on precipitation, spatial and temporal variations in atmospheric temperature, pressure and humidity, physical features of the earth, factors influencing the distribution and activities of man etc. This information would be helpful in fish/shrimp farming activities such as pond construction and assessment of suitability of pond.
- Every year, life and property are lost due to climate-related hazards such as tropical cyclones, rough seas, lighting etc. This course gives adequate awareness about all these climate- related natural disasters. This course also deals with action of nature upon man and reaction or action of man

upon nature. So that the students will get precise information about greenhouse gases, greenhouse effect, global warming and their influence on fisheries. From this course, students also get information on weather forecasting and effects of weather and climate changes on fisheries.

Chapter-1

Nature of Atmosphere

Introduction to Meteorology

The earth is surrounded by a large envelope of gases called atmosphere. The condition of the atmosphere and the changes it undergoes influence all physical, chemical and biological activities over the earth. The study pertaining to lower atmosphere and its changes is called atmosphere physics or meteorology. In modern usage, meteorology denotes the science of weather and includes the study of atmospheric phenomena.

Definition: Meteorology can be defined as the science that deals with the study of the atmosphere.

Divisions of Meteorology

- **Marine meteorology**: Meteorology pertaining to the sea and ocean is called marine meteorology.
- **Aeronomy**: The science of aeronomy investigates both the chemical and physical properties of the atmosphere and the changes that occur in these as the result of external and internal forces.
- **Bio-meteorology**: The scientific study of the interactions between living organisms and their atmospheric environment called biometeorology.

Difference Between Weather and Climate

	Weather	Climate
Definition	Short-term atmospheric conditions in a specific place at a specific time.	The long-term average of weather patterns in a region over extended periods (typically 30 years or more).
Duration	Minutes to weeks.	Decades to Centuries
Elements	Temperature, humidity, precipitation, wind, visibility, etc.	Average temperature, average rainfall, seasonal patterns, frequency of extreme events, etc.
Changes	It can change rapidly from day to day or hour to hour.	Changes are gradual and are observed over long durations.
Predictability	Forecasted using meteorological instruments for short-term predictions.	Predicted based on long-term data and patterns.
Impact	Immediate effects on daily activities, travel, and events.	Affects ecosystems, agriculture, infrastructure, and overall living conditions in a region.
Example:	Today's temperature, this week's rainfall, tomorrow's wind speed.	The Mediterranean has a warm, dry summer and a mild, wet winter. The Sahara has a hot, dry climate year-round.

Composition of Atmosphere

- ❑ The atmosphere is described as the air that surrounds the earth
- ❑ The thickness of the earth's atmosphere is about 480 km. 99 percent of the thickness lies up to the height of 32 km from the earth
- ❑ With increasing altitude, the air pressure decreases
- ❑ The atmosphere has a mixture of gases that sustains life on earth
- ❑ The earth's gravity helps hold the atmosphere in place
- ❑ The major role of the atmosphere is to contain the entry of ultraviolet rays

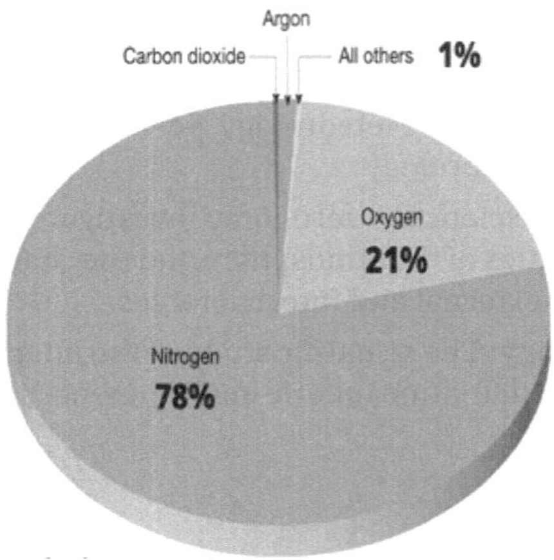

Composition of earth's atmosphere

As per NASA, the composition of the earth's atmosphere is as mentioned below

- **Nitrogen** - 78%
- **Oxygen** - 21%
- **Argon** - 0.93%
- **Carbon dioxide** - 0.04%
- Trace amounts of neon, helium, methane, krypton and hydrogen, as well as water vapour.

The atmosphere, comprising layers of gas, surrounds a planet and is held in place by the gravitational force of the planetary body. A planet retains its atmosphere when it possesses sufficient gravity and maintains low atmospheric temperatures.

- Earth's atmosphere primarily consists of nitrogen (78%), oxygen (21%), argon (0.9%), carbon dioxide (0.04%), and trace gases. Additionally, a variable amount of water vapor is present (approximately 1% at sea level), decreasing with altitude.
- Carbon dioxide is a significant contributor to the greenhouse effect. While transparent to incoming solar radiation, it becomes opaque to outgoing terrestrial radiation. It absorbs a portion of terrestrial radiation and reflects some back toward earth's surface.
- Dust particles, originating from various sources such as fine soil, smoke-soot, pollen, and disintegrated meteor particles, are also found in the atmosphere. These particles, along with salt particles, act as hygroscopic nuclei around which water vapor condenses, forming clouds.

Ozone Gas

- Present around 10-50 km above the earth's surface and acts as a sieve, absorbing UV (ultraviolet rays) from the sun.

Water Vapour

- Water vapour is a variable gas, declines with altitude.
- It also drops towards the poles from the equator.
- It acts like a blanket letting the earth from becoming neither too hot nor too cold.
- It also contributes to the stability and instability in the air.

Dust Particles

- Dust particles are in higher concentrations in temperate and subtropical regions due to dry winds in contrast to the polar and equatorial regions.
- They act as hygroscopic nuclei over which water vapour of the atmosphere condenses to create clouds.

Nitrogen

- The atmosphere is composed of 78% nitrogen.
- Nitrogen cannot be directly utilized from the air.
- Living organisms require nitrogen to produce proteins.
- The Nitrogen cycle is the process by which necessary nitrogen is supplied to living things.

Oxygen

- The atmosphere is composed of 21% oxygen.
- Oxygen is utilized by all living things and is essential for respiration.
- It is necessary for combustion.

Argon

- The atmosphere is composed of 0.9% argon.
- It is mainly used in light bulbs.

Carbon Dioxide

- The atmosphere is composed of 0.03% carbon dioxide.
- Plants use it to produce oxygen.
- It is significant as it is opaque to outgoing terrestrial radiation and transparent to incoming solar radiation.
- It is also one of the gases responsible for the greenhouse effect.

Structure of the atmosphere

There are five layers in the structure of the atmosphere depending upon temperature. These layers are:

- Troposphere
- Stratosphere
- Mesosphere
- Thermosphere
- Exosphere

Layers of Atmosphere

(i) Troposphere

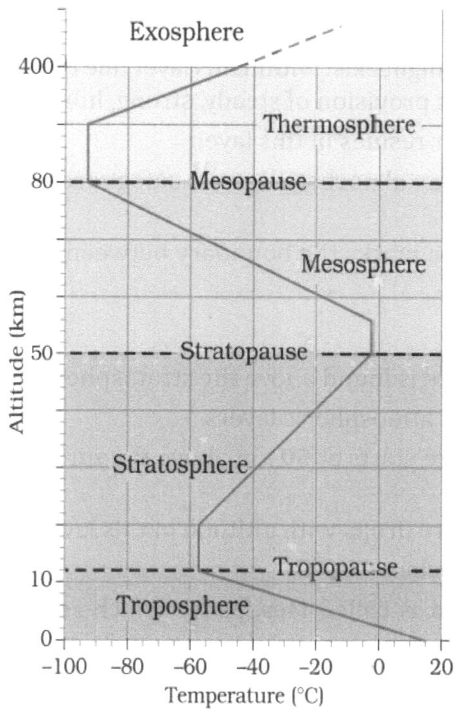

- It is considered the lowest layer of earth's atmosphere.
- The troposphere extends from the earth's surface up to a height of 8 km (at the poles) to 18 km (at the equator).
- The higher height at the equator is primarily due to the presence of hot convection currents that push gases upward.
- All types of weather changes occur within this layer.
- This layer contains water vapor and various particles.
- Temperature decreases with increasing altitude in the atmosphere at a rate of 1 degree celsius for every 165 meters of height. This is known as the normal lapse rate.
- The tropopause, a transitional zone, separates the troposphere from the stratosphere.

(ii) Stratosphere

- It is the second layer of the atmosphere located above the troposphere.
- It extends up to a height of 50 km from the earth's surface.
- This layer is notably dry, containing minimal water vapor.
- Advantages for flight exist within this layer due to its position above stormy weather and its provision of steady, strong, horizontal winds.
- The ozone layer resides in this layer.
- The ozone layer absorbs UV rays, safeguarding earth from harmful radiation.
- The stratopause marks the boundary between the Stratosphere and the Mesosphere.

(iii) Mesosphere

- The Mesosphere is found above the stratosphere.
- It is the coldest atmospheric layers.
- The mesosphere starts at 50 km above the surface of the earth and goes up to 80 km.
- The temperature drops with altitude in this layer.
- By 80 km it reaches -100°C.
- The upper limit is called Mesopause which separates Mesosphere and Thermosphere.

(iv) Thermosphere

- This layer is found above Mesopause from 80 to 400 km.
- Radio waves that are transmitted from the earth are reflected by this layer.
- The temperature starts increasing again with increasing height in this layer.
- Aurora and satellites occur in this layer.

(v) Ionosphere

- The lower thermosphere is called the Ionosphere.
- The ionosphere consists of electrically charged particles known as ions.
- This layer is defined as the layer of the atmosphere of earth that is ionized by cosmic and solar radiation.
- It is positioned between 80 and 400 km above the Mesopause.

(vi) Exosphere

- It is the outermost layer of the atmosphere.
- The zone where molecules and atoms escape into space is mentioned as the exosphere.
- It extends from the top of the thermosphere up to 10,000 km.

Heat Energy of Atmosphere

The Earth's atmosphere is essential for maintaining our planet's temperature and supporting life. Heat energy, primarily originating from the Sun, drives various atmospheric processes crucial for weather patterns and climate dynamics. Solar radiation, including visible light, UV, and infrared radiation, is absorbed by the atmosphere, heating its components such as the earth's surface, clouds, and gases. This absorption leads to the transfer of heat energy through mechanisms like radiation, conduction, convection, and advection. For instance, warm air rises and cool air sinks, creating vertical movements known as convection, while horizontal movements, like wind and ocean currents, facilitate the transfer of heat across different regions. Additionally, greenhouse gases like carbon dioxide and water vapor trap heat, helping regulate Earth's temperature.

Process of Heat Transmission

There are four common types of energy transfer involved in heating the atmosphere. They are Radiation, Conduction, Convection and Advection.

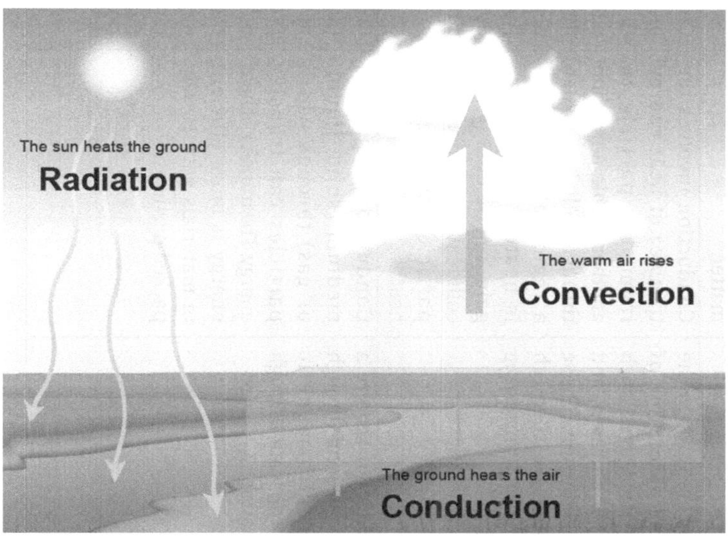

Property	Radiation	Conduction	Convection	Advection
Definition	Transfer of heat energy through electromagnetic waves, without the need for a medium.	Transfer of heat energy through direct contact between particles of matter.	Transfer of heat energy through the movement of fluid (liquid or gas) particles.	Horizontal movement of air or water masses carrying heat energy.
Mechanism	Radiation occurs through the emission and absorption of electromagnetic waves, such as visible light, infrared, and ultraviolet radiation. These waves propagate through space and matter, transferring heat energy.	Conduction involves the transfer of heat energy from one particle to another within a medium through collisions. When a region of a material is heated, its particles gain kinetic energy and collide with neighbouring particles, transferring heat.	Convection relies on the movement of fluid particles to transfer heat energy. In natural convection, warmer fluid expands, becomes less dense, and rises, while cooler fluid contracts, becomes denser, and sinks, creating circulation currents.	Advection occurs when a fluid, typically air or water, carries heat energy horizontally from one location to another. This movement is driven by pressure gradients, temperature differences, or other forces, transporting heat over large distances.
Medium	Radiation does not require a medium and can occur through vacuum. Electromagnetic waves can propagate through space and matter.	Conduction requires a medium (solid, liquid, or gas) through which particles can transfer energy. The transfer of heat energy is more efficient in materials with closely packed particles.	Convection occurs in fluids (liquids and gases), where the movement of fluid particles facilitates the transfer of heat. The rate of heat transfer through convection depends on fluid velocity, temperature gradients, and fluid properties.	Advection typically occurs in large-scale fluid movements, such as wind in the atmosphere or ocean currents. The horizontal movement of fluids carries heat energy from one region to another, influencing regional climates and weather patterns.

Property	Radiation	Conduction	Convection	Advection
Example	Examples of radiation include the sun radiating heat energy through space, which is absorbed by the earth's atmosphere and surface. Other examples include infrared radiation emitted by heated objects and microwave radiation used in cooking.	Examples of conduction include heating a metal rod at one end, causing the heat to conduct along its length and warm the entire rod. Conduction also occurs when touching a hot object, where heat is transferred directly to the skin.	Examples of convection include boiling water in a pot, where heated water rises, transfers heat to the cooler water above, and creates circulation currents. Natural convection in the atmosphere contributes to weather phenomena such as thunderstorms and sea breezes.	Examples of advection include warm ocean currents like the Gulf Stream, which transport heat from the tropics toward higher latitudes, influencing regional climates and weather patterns. Atmospheric advection occurs when warm or cold air masses move horizontally, affecting temperature distributions and weather conditions.

Heating of the Atmosphere

The atmosphere does not absorb short wave radiation. Therefore, most of the radiation is transmitted to earth without heating the atmosphere. The insolation chiefly heats up the earth's surface. The heated surface of the earth reflects back the heat to atmosphere in the form of long-waves (terrestrial radiation). Insolation is received only during the day-time. But transmission of terrestrial radiation takes place all through the day and night. As a result of this phenomenon nights are cooler than the day.

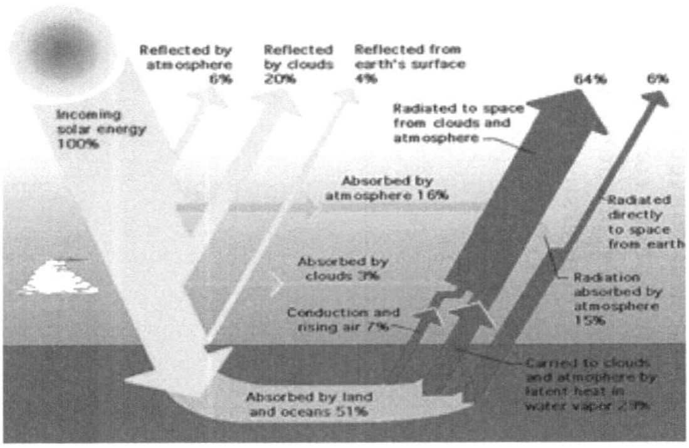

Heat Budget of the Earth

The average temperature of the earth generally does not change. The amount of heat reflected back by earth through terrestrial radiation equals the amounts of insolation or incoming solar radiation. In the way there is balance between insolation and terrestrial radiation. This balance is known as the heat budget of the earth.

Disposal of Insolation

(i) Insolation

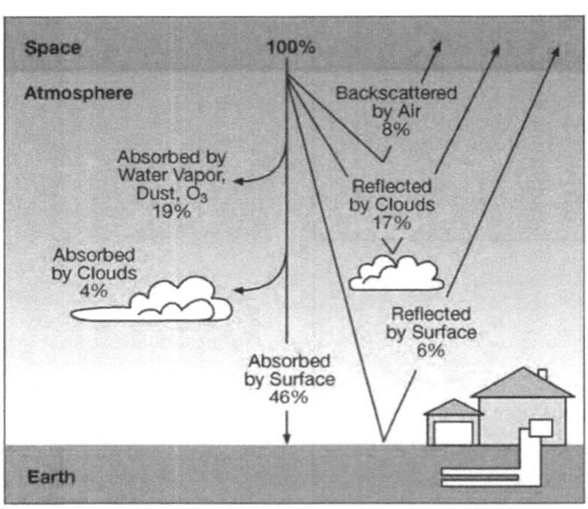

The energy for all atmospheric processes is derived from the sun. The sun radiates energy at a temperature of 6000°C. Incidence of the solar radiation on the earth's surface is referred as insolation.

The sun's rays suffer depletion due to absorption, scattering and reflection and by the atmospheric constituents, particulates or aerosols and clouds. Most of the ultraviolet radiation is absorbed by the ozone layer. Part of the radiation is reflected back to space by water, snow and deserts and the rest is absorbed by land and water. Nearly 19 percent of the incoming solar radiation is absorbed in the atmosphere and 34 percent scattered or reflected back to space. The rest 47 percent of the radiant energy is absorbed by the earth and transformed into heat. Heating extends to considerable depth in water due to turbulent mixing and overturning. Oceans store energy for long periods. Over land, however, the heating effect is restricted to a few centimeters. Heat is transferred upwards by conduction and eddy transfer. Large masses of heated air become buoyant and are carried upward by convection. Cooling takes place in the vertical uplift and water vapor condenses into liquid water drops, liberating latent heat. The amount of insolation received at a place on the earth at any time depends on,

- Solar output, i.e. energy emitted from the sun.
- Distance of the earth from the sun.
- Transparency of the atmosphere.
- Period of insolation.
- Angle at which sun's rays strike the earth.
- Composition of the earth's surface.

(ii) Albedo

The term albedo denotes the reflectivity for sunlight of a rough (diffusely reflecting) surface or of an entire planet. It is defined as the ratio of diffusely reflected radiation to the incident radiation.Thus, an albedo of 1(or 100 percent) would apply to a perfectly white diffuse reflector, while an albedo of zero (or 0 percent) would apply to a perfectly black body. The average or overall albedo of a planet determines its mean temperature.

Irregular Heating of Atmosphere

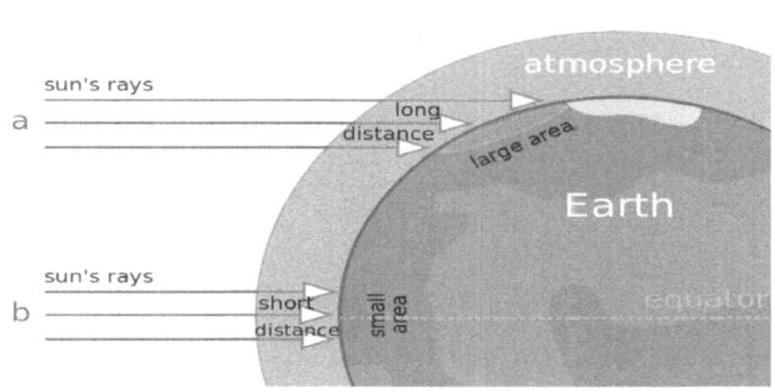

Uneven heating of the planet also occurs due to the Earth's curvature. The majority of energy is received near the equator, where the Sun's rays directly (or nearly directly) hit the planet. As one moves towards the poles, the Sun's energy becomes more diffused and spreads out over a larger area. Ultimately, the equator receives more energy from the Sun than it can radiate back into space, resulting in a surplus of energy. Conversely, at the poles, the planet radiates more energy out into space than it receives from the Sun, leading to a deficit of energy in these regions. The primary function of weather is to transfer the surplus energy and heat to the poles while bringing colder air towards the equator to establish equilibrium.

Chapter-2

Temperature

- The degree of hotness or coldness is known as temperature.
- When any object receives energy in the form of heat, its temperature increases.
- Air or atmosphere receives the heat energy from the sun and its temperature increases.
- Due to different amount of heat energy receipt at different places, the air temperatures at different places also vary.
- The variations in air temperatures basically result into air motions so as to equalize the energy content of the different regions of the earth.

Thus, temperature of air can be regarded as the basic cause for weather changes.

A. Instruments Used for Temperature Measurements

(i) Maximum Minimum Thermometer

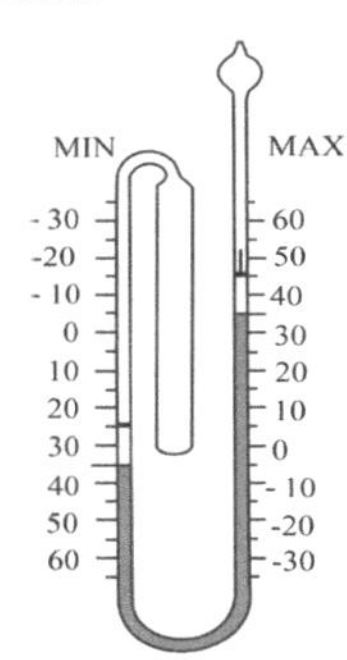

- It helps to measure the maximum and minimum temperature in a day of a place.
- It is a U-shaped tube filled with alcohol and mercury.
- The right tube gives us the maximum temperature while the left tube gives us the minimum temperature.
- When the temperature increases, the alcohol in the left tube expands. This causes the mercury to move

towards the right tube. A metal index on the right tube also shifts which gives us the maximum temperature for that day.

- When the temperature decreases, the alcohol contracts in the left tube and the mercury moves into the left tube. A metal index in the left tube shift and give us the minimum temperature of the day.

(ii) Thermograph

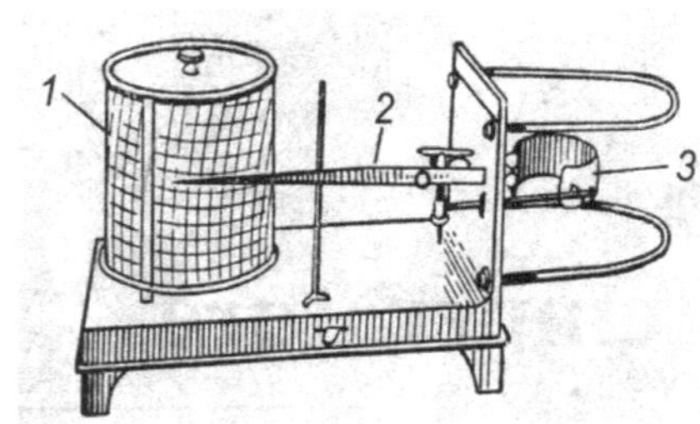

This recording thermometer is not often seen abroad ship. A pen, attached to a metallic coil which expands and contracts, records the temperatures on a drum moved by clockwork. A commonly used thermograph is one which uses a bimetallic strip as reactor. Two thin, curved sheets of metal of widely different thermal expansion are welded together. When the temperature changes, the two metals will expand unequally, and the curvature of the strip will change. This change is transferred through levers to a pen that traces the temperatures on the thermogram.

(iii) Bimetallic Thermometer

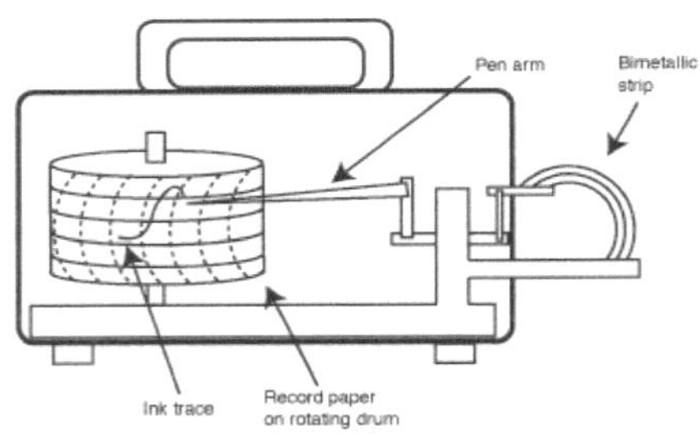

The bimetallic thermometer uses the bimetallic strip which converts the temperature into the mechanical displacement. The working of the bimetallic strip depends on the thermal expansion property of the metal. The thermal expansion is the tendency of metal in which the volume of metal changes with the variation in temperature. Every metal has a different temperature coefficient. The temperature coefficient shows the relation between the change in the physical dimension of metal and the temperature that causes it. The expansion or contraction of metal depends on the temperature coefficient, i.e., at the same temperature the metals have different changes in the physical dimension.

Working Principle of Bimetallic Thermometer

- The working principle of bimetallic thermometer depends on the two fundamental properties of the metal.
- The metal has the property of thermal expansion, i.e., the metal expands and contract concerning the temperature.
- The temperature coefficient of all the metal is not same. The expansion or contraction of metals is different at the same temperature.

Difference between heat and temperature

- The interaction of incoming solar radiation (insolation) with the atmosphere and the earth's surface creates heat. This is measured in terms of temperature.
- While heat represents the molecular movement of particles comprising a substance, the temperature is the measurement in degrees of how hot (or cold) a thing (or a place) is.
- So, heat is the energy which makes things or objects hot, while temperature measures the intensity of heat (degree of hotness/coldness).
- Heat and temperature are closely related to each other because gain or loss of heat is necessary to raise or lower the temperature
- The Celsius scale is the internationally accepted scale for reporting air temperature.

B. Temperature Variation

Air temperature at any location is changes during a day, week, month, and year or for any period. On this basis it is classified as-

1. Periodic variation.
   i. Annual temperature variation or Annual temperature cycle.
   ii Diurnal temperature variation or Daily temperature cycle.
2. Horizontal variation
3. Vertical Variation.

1. Periodic Temperature Variation

The temperature continuously changes during a day, week, month, year or any period and this change is called periodic temperature variation. Periodic temperature variations are –

(i) Annual temperature variation or Annual temperature cycle

- The annual temperature variation gives rise to seasons i.e. summer and winter.
- The annual temperature range varies greatly from place to place.
- It reflects the daily increase in insulation from mid-winter to mid-summer and decrease in the same from mid-summer to mid-winter summer and decrease in the same from mid-summer to mid-winter usually there is a temperature lag of 30 to 40 days after the period of maximum and minimum insulation.
- In the northern hemisphere winter minimum occurs in January and summer maximum in July and vice versa in the southern hemisphere.
- The smaller range occurs near equator and largest in high latitudes.

- The difference between the highest and lowest temperature for a given period is known as temperature range.
- In the northern hemisphere it is summer from 21st of March to 22nd of September and winter from 23rd September to 20th March and vice-versa in southern hemisphere.

(ii) Diurnal temperature variation or Daily temperature cycle

- The minimum air temperature occurs at about sunrise, after which there is a constant rise till it reaches to maximum.
- The maximum air temperature is recorded between 1300 hrs and 1400 hrs although the maximum solar radiation is reaches at the moon.
- A steady fall in temperature untill sunrise is noticed after is attains maximum. Thus through out the day in March displays one maximum and one minimum. The difference between the two is called the diurnal range of air temperature.
- The diurnal range of air temperature is more on clear days while cloudy weather sharply reduces daily amplitudes.
- The diurnal range of temperature is also influenced by soils and their coverage in addition to seasons.
- Addition of daily maximum and minimum temperature divided by two is nothing but daily mean / average temperature.
- In northern hemisphere winter minimum occurs in January and summer maximum in July.

2. Horizontal Temperature Variation

- The rate of change of temperature with a horizontal distance is known as Temperature Gradient.
- On an average, temperature decreases from the equator towards the poles and thus low latitudes are characterized by highest temperature whereas high latitudes record lowest temperature.
- It may be noted that highest temperature is never recorded at the equator instead it is recorded near the tropics of capricorn and cancer because a sizeable portion of incoming solar radiation is reflected by clouds and a large part of heat is spent in the process of evaporation near the equator.
- On the other hand, most of the incoming solar radiation reaches the ground surface due to clear sky near the tropics of cancer and capricorn.
- Normally, temperature decreases from equator towards the poles.
- This change of temperature rather decrease of temperature pole-ward is called temperature gradient.
- The rate of decrease of temperature between the tropics of cancer and capricorn is rather low and therefore temperature gradient is insignificant but temperature decreases more rapidly from the tropics pole-ward and

therefore temperature gradient is also very high. The horizontal distribution of temperature is represented and studied with the help of isotherms.

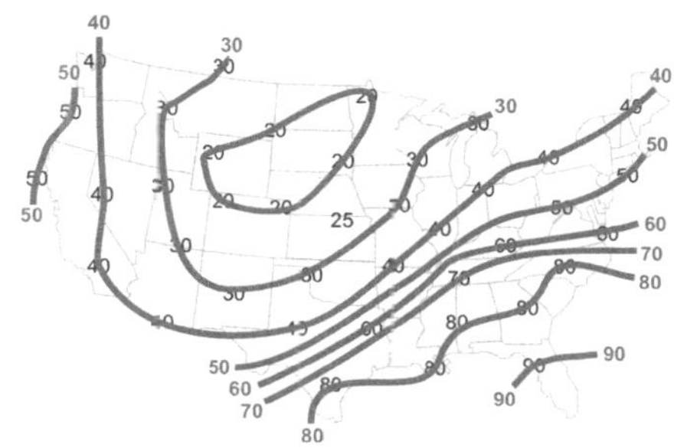

Isotherm

Isotherm is defined as the line on the weather map joining the places of equal temperature.

3. Vertical Temperature Variation

- Vertical temperature variation does not show uniform behaviour and the atmosphere can be divided into four spheres.
- Troposphere - Temperature decreases from 15^0C at earth surface up to -60^0C at 11 km height.
- Stratosphere - Temperature increases from -60 ^{0}C to 0 ^{0} C at 50km height.
- Mesosphere - Temperature fall and reaches about -90 ^{0}C at 80 km height.
- Thermosphere- Temperature increases due to absorption of solar radiation by atomic oxygen, up to 950^0C at 350 km height and 1700^0C at undefined upper Limit.

Temperature inversion

Certain special conditions in the atmosphere at some altitude may produce a reversal of the normal lapse rate, so that the temperature actually increases with an increase in altitude. This is known as temperature inversion. This can occur only when a warm layer of air overlies a colder layer. The layer in which temperature increases with altitude is called inversion layer.

Actual lapse rate

The actual decrease in temperature with elevation at a given place is called actual lapse rate. It is determined by actually observing the temperature at different elevations. It changes from place to place and time to time.

Normal lapse rate

The average rate of temperature decrease with elevation is called "Normal lapse rate". This is determined by taking the average of the different lapse rate observed at different places and different times. Normal lapse rate is 6.5°C.

C. Effects of Vertical Air Motion on Temperature

Vertical air motion plays a significant role in determining temperature patterns in the atmosphere.

(i) Adiabatic Heating and Cooling

- As air rises vertically, it expands due to decreasing atmospheric pressure. This expansion leads to adiabatic cooling, where the air temperature decreases with height.
- Conversely, as air descends, it compresses due to increasing atmospheric pressure. This compression causes adiabatic heating, resulting in an increase in temperature with descent.
- These adiabatic processes are fundamental to understanding temperature changes associated with vertical air motion.

(ii) Cloud Formation

Vertical air motion is closely tied to cloud formation processes. When moist air rises, it cools adiabatically. If it reaches its dew point, condensation occurs, leading to cloud formation. Clouds can influence local temperature by reflecting sunlight during the day and trapping heat at night.

(iii) Precipitation

Vertical air motion is often associated with precipitation. Rising air cools and may reach its saturation point, leading to the formation of clouds and subsequent precipitation. This can have a cooling effect on the surrounding air due to evaporative cooling during rainfall.

(iv) Temperature Inversion

In some situations, especially during stable atmospheric conditions, vertical air motion can result in temperature inversions. Inversions occur when a layer of warm air traps cooler air beneath it, leading to a decrease in temperature with height instead of the typical increase. These inversions can have significant impacts on air quality and weather patterns.

(v) Localized Temperature Variations

- Vertical air motion can create localized temperature variations. For example, air moving downslope (downslope winds) can lead to warming due to compression, resulting in warmer temperatures at lower elevations (e.g., foehn winds).
- Conversely, updrafts associated with convection can transport warm air aloft, leading to cooler surface temperatures in affected areas.

(vi) Vertical Temperature Profiles

- Vertical air motion influences the vertical temperature profile of the atmosphere. In unstable conditions, where warm air rises and cold air descends, a steeper lapse rate (rate of temperature decrease with height) is observed. In stable conditions, where air parcels resist vertical displacement, a shallower lapse rate is observed.

Chapter--3

Humidity and Water Vapour

The concentration of water vapour present in the air is known as humidity. The widely employed primary measurements of humidity are absolute, relative, and specific humidity.

Types of Humidity

(i) Absolute Humidity

Absolute humidity describes the water content present in the air and is expressed in either gram per cubic meter or grams per kilogram. The absolute humidity in the atmosphere ranges from near zero to roughly 30 grams per cubic metre.

Mathematically, absolute humidity is defined as the mass of the water vapour divided by the volume of the air and water mixture, which is expressed as

$$\square H = \frac{mH_2O}{V_{net}}$$

Where,

mH_2O = Mass of the water vapour

V_{net} = Volume of air and water mixture

The absolute humidity varies with respect to air temperature, and pressure changes if the volume is not fixed. The mass of water vapour per unit volume in the equation is also defined as volumetric humidity.

(ii) Relative Humidity

Relative humidity, expressed as a percentage, signifies the ratio of the partial pressure of water vapor in an air-water mixture to the equilibrium water vapor pressure over a flat surface of pure water at a specific temperature. It serves as a vital measure in weather forecasts, indicating the potential for precipitation, dew formation, or fog. During hot summer conditions, elevated relative humidity impedes the evaporation of sweat from the skin, thereby raising human body temperature.

(iii) Specific Humidity

The ratio of water vapour's mass to the air parcel's total mass is known as specific humidity.

Water Vapour

- Water vapour is water in gaseous state instead of liquid form. It is the outcome of either, the process of evaporation or sublimation.
- It is very significant to the weather and climate.
- It is one of the earth's most vital greenhouse gas.
- It accounts for about 90%of the earth's natural greenhouse effect, which enables keep the earth warm enough to support life.
- When water evaporates from the surface of the earth, it cools the surface. The process of cooling of water vapour keeps the surface from getting too hot.
- It is rarely referred to the global warming debate that the surface cooling effect of evaporation is more powerful than its greenhouse warming effect.
- Water vapor is water in its gaseous state-instead of liquid or solid. It is totally invisible.
- When the air gets saturated with the water vapour, at a specific temperature it will form frost as it starts to condense.
- Water vapour holds its importance as evidence of the existence of extraterrestrial liquid water on a planet.

Relationship Between Temperature and Humidity

Relative humidity and temperature: Relative humidity (RH) is a measure of the amount of moisture in the air compared to the maximum amount of moisture the air can hold at a given temperature. Warm air can hold more moisture than cool air. Therefore, as temperature increases, the air's capacity to hold moisture also increases, and vice versa. If the amount of moisture in the air remains constant and the temperature increases, the relative humidity decreases because the air can hold more moisture.

Dew Point and Temperature: Dew point is the temperature at which the air becomes saturated with water vapor and starts to condense into liquid water (dew). It's a more accurate measure of the moisture content in the air than relative humidity. When the temperature and dew point are close together, the air feels humid; when they're far apart, it feels dry. If the air temperature drops to the dew point, condensation occurs, leading to the formation of dew, fog, or clouds.

Heat Index: The heat index combines temperature and humidity to determine how hot the air feels to the human body. High humidity levels can make hot temperatures feel even hotter because the body's ability to cool itself through sweating is impaired. Therefore, a high heat index, resulting from both high temperature and high humidity, can increase the risk of heat-related illnesses such as heat exhaustion and heatstroke.

Evaporation and cooling: Humidity affects the rate of evaporation, which in turn influences cooling processes. In areas with low humidity, such as deserts, sweat evaporates quickly from the skin, providing a cooling effect. Conversely, in humid environments, evaporation is slower, making it harder for the body to dissipate heat, leading to discomfort in high temperatures.

Weather patterns: Temperature and humidity variations play a significant role in weather patterns. High humidity levels can contribute to the formation of clouds and precipitation, leading to rainy or stormy weather conditions. On the other hand, low humidity levels can result in clear skies and dry weather.

Example

Shrimp Farming in Coastal Areas

Shrimp farming, particularly in coastal regions, is heavily influenced by temperature and humidity conditions. Shrimp farming ponds require specific environmental conditions for optimal shrimp growth and health, including water temperature and salinity levels, which are closely linked to temperature and humidity.

Temperature Influence: High temperatures combined with high humidity levels can lead to elevated water temperatures in shrimp farming ponds. Warm water temperatures can accelerate the growth rates of shrimp and promote higher feed conversion efficiency. However, excessive temperatures, especially in conjunction with high humidity, can also increase the risk of disease outbreaks and reduce oxygen levels in the water, negatively impacting shrimp health.

Humidity Influence: Humidity levels can affect water evaporation rates in shrimp farming ponds. In regions with low humidity, such as arid coastal areas, water evaporation rates may be higher, leading to increased salinity levels in shrimp ponds. Elevated salinity levels can stress shrimp and impact their growth and survival rates. Managing humidity levels and water quality parameters is essential for maintaining optimal conditions for shrimp farming and ensuring sustainable production.

Distribution of Water Vapour in Atmosphere

Vertical Distribution

Troposphere: The troposphere is the lowest layer of the atmosphere, extending from the earth's surface up to about 8-15 kilometres (5-9 miles) in altitude, depending on location and season. It contains the majority of the Earth's weather systems and atmospheric water vapor. The concentration of water vapor is highest near the Earth's surface and decreases with altitude. This decrease occurs primarily due to decreasing temperature with height, as cooler air can hold less water vapor than warmer air.

Stratosphere: Above the troposphere lies the stratosphere, which extends from the tropopause (the boundary between the troposphere and the stratosphere) to approximately 50 kilometres (31 miles) in altitude. While some water vapor exists in the lower stratosphere, its concentration is relatively low compared to the troposphere. Water vapor concentrations diminish rapidly with altitude in the stratosphere due to the lack of significant water sources and the extremely low temperatures, which inhibit the presence of gaseous water vapor.

Horizontal Distribution

Latitude: Water vapor concentrations vary with latitude. Near the equator, where temperatures are higher, water vapor concentrations tend to be higher as well. This is because warmer air can hold more moisture. Conversely, at higher latitudes, especially towards the poles, water vapor concentrations are lower due to colder temperatures, which reduce the capacity of the air to hold moisture.

Regional Variation: The distribution of water vapor also varies regionally due to factors such as proximity to water bodies, prevailing wind patterns, and local weather systems. Coastal areas and regions with high humidity levels tend to have higher concentrations of water vapor compared to arid inland regions. Wind patterns can transport moisture inland from oceanic sources, leading to higher humidity levels in coastal areas.

Weather Systems: Weather systems such as cyclones, hurricanes, and fronts play a significant role in transporting moisture across different regions. These systems can create localized areas of high humidity and precipitation. For example, the leading edge of a warm front typically brings warm, moist air into an area, leading to increased humidity levels and potentially precipitation. Similarly, the circulation within cyclones and hurricanes can transport moisture over large distances, affecting the distribution of water vapor in the atmosphere.

Humidity Instruments and Measurements

Hygrometer

Hygrometers are instruments that indirectly gauge humidity by detecting alterations in the physical or electrical characteristics of materials caused by moisture absorption. Materials like hair, skin, membranes, and thin strips of wood experience dimensional changes as they absorb water, and these alterations are directly proportional to humidity levels. These devices are employed for measuring relative humidity within the range of 20 to 90 percent, with an accuracy of approximately ± 5 percent. They are suitable for operation within temperature ranges below 70°C.

Psychrometer

Psychrometers work on the principle of evaporative cooling. They consist of two thermometers: a dry bulb thermometer and a wet bulb thermometer. The wet bulb is covered with a moistened cloth or wick. As the moisture evaporates from the wet bulb, it causes cooling, and the temperature difference between the dry and wet bulb thermometers is used to calculate relative humidity.

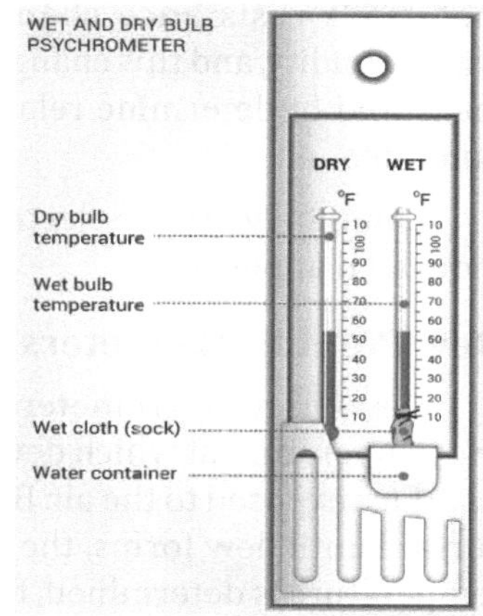

Hair Hygrometer

Hair hygrometer is the simplest and oldest type of hygrometer. It is made using hair as shown in Figure. Human hair lengthens by 3 percent when the humidity changes from 0 to 100 percent, the change in length can be used to control a pointer for visual readings or a transducer such as a linear variable differential transformer (LVDT) for an electrical output. The hair hygrometer has an accuracy of about ±5 percent for the humidity range 20 to 90 percent over the temperature range 5 to 40°C.

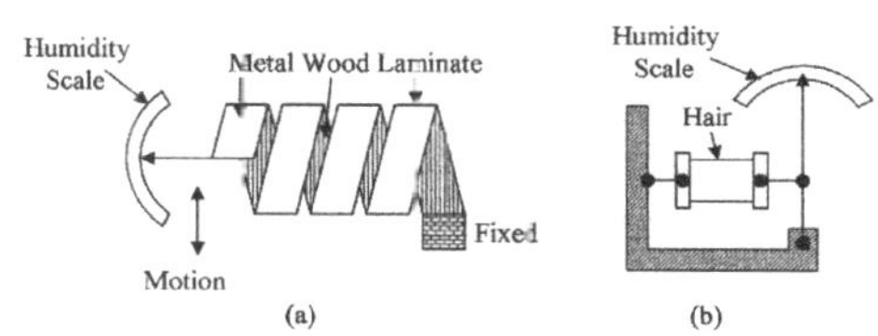

Capacitive Hygrometers

- Capacitive hygrometers use a humidity-sensitive material whose dielectric constant changes with humidity. This change in dielectric constant alters the capacitance of the sensor, which is then measured to determine relative humidity.

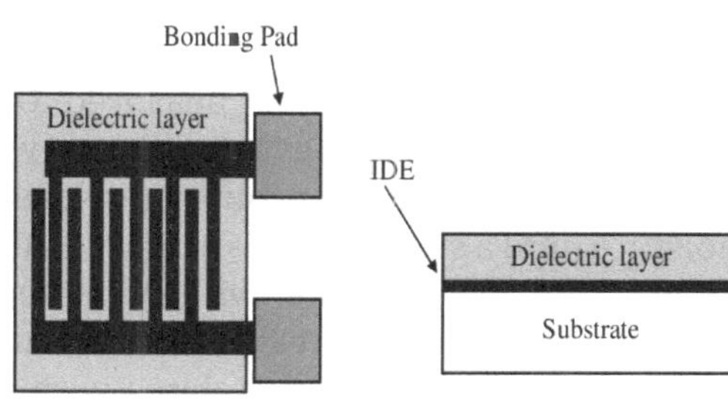

- Widely used in electronic devices, weather stations, and industrial applications due to their accuracy and fast response time.

Resistive Hygrometers

Resistive hygrometers use a humidity-sensitive material, such as a polymer or salt, between two conductive electrodes. The material's resistance changes with humidity, and this change is measured to determine relative humidity.

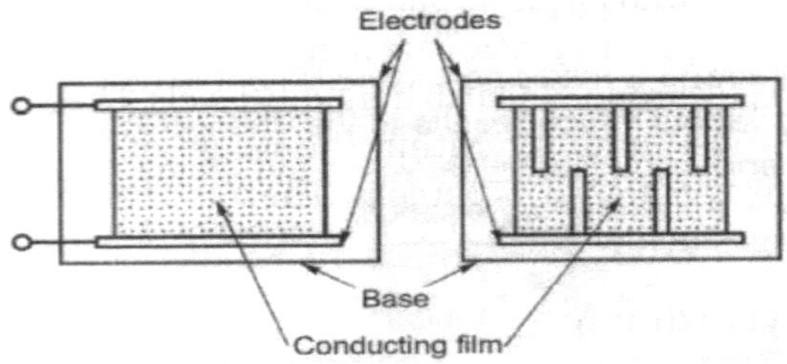

Resistive Hygrometer

Commonly used in electronic devices, environmental monitoring systems, and industrial applications.

Dew Point Hygrometers

Dew point hygrometers measure the temperature at which dew forms on a surface exposed to the air. By cooling a surface until dew forms, the dew point temperature is determined, from which relative humidity can be calculated.

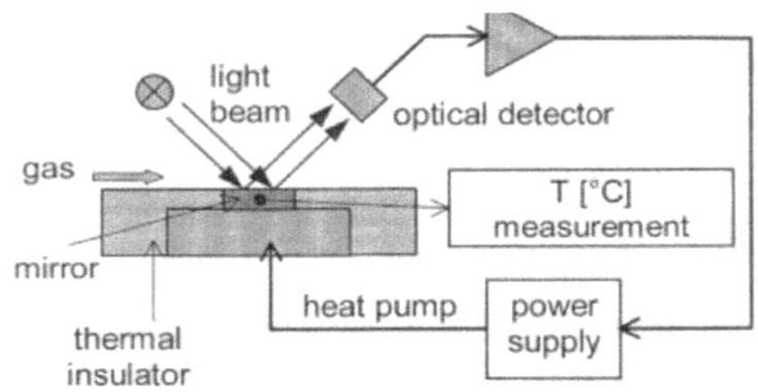

Used in industrial drying processes, food production, and meteorology for precise humidity measurements.

Gravimetric Hygrometers:

A gravimetric hygrometer operates on the principle of measuring humidity by quantifying the change in mass of a hygroscopic material as it absorbs or desorbs water vapor from the surrounding air. Initially, a known mass of the hygroscopic material is weighed while dry, and then exposed to the air whose humidity is to be measured. Over time, the material reaches equilibrium with the surrounding air, absorbing or releasing moisture until a stable mass is achieved. After equilibration, the container with the material is weighed again to determine the change in mass, which directly correlates with the humidity of the air. This change is then used to calculate the relative humidity of the air, providing a precise measurement of moisture content. Gravimetric hygrometers offer high accuracy and stability, particularly at low humidity levels, but may require careful calibration and handling.

Hygrograph

The hygrometer may be transformed into a self-recording instrument, and it is then called a hygrograph

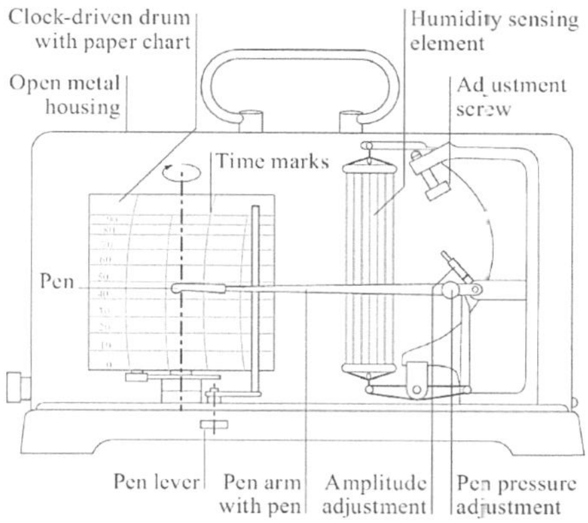

Chapter-4

Condensation and Precipitation

Condensation is the process through which the physical state of matter changes from the gaseous phase into the liquid phase. For example, condensation occurs when water vapour (gaseous form) in the air changes into liquid water when it comes in contact with a cooler surface. When the water in the air comes in contact with a cold surface, it condenses to form water droplets.

Process of Condensation

Condensation of water happens when water changes its phase from gaseous state to liquid or crystal shape. At high pressure and low temperature, any gas can condense. Technically, the process of condensation can happen at any temperature as long as the pressure of the liquid state of the gas is less than the pressure of the condensing gas. The molecules in the matter slow down during the process of condensation because the heat energy is which causes a change within the three states of matter, ultimately transforming the matter into the solid-state.

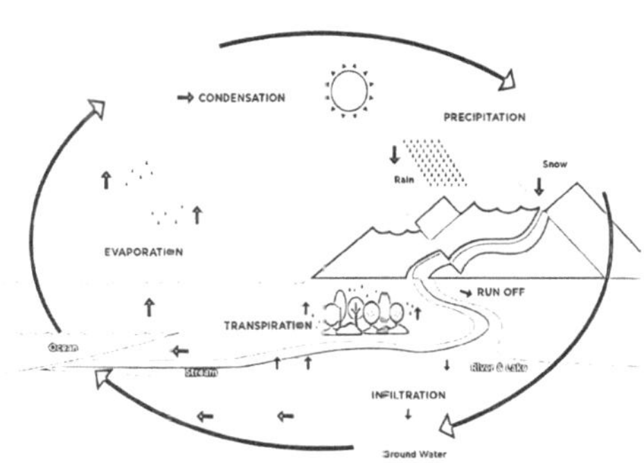

Condensation – Water Cycle

- ❑ Condensation is important to the water cycle as it is responsible for the formation of clouds.

- Water vapour present in the air is responsible for the formation of clouds which ultimately comes down in the form of rain.
- This phase change of water between solid, liquid and gas are because of the movement of water molecules.
- In vapour form, water molecules are arranged randomly as compared to the liquid state.
- As condensation happens, water molecules become more organized and as a result, heat is released into the atmosphere leading to a change of phase from the vapour state to the liquid state.
- This generally occurs in the atmosphere when warm air rises up and cools down. For condensation to take place, it is very important that the atmosphere is fully saturated (to reach maximum vapour pressure). Usually, condensation takes place around dust particles or smoke or microscopic bacteria. It plays a very significant role in the water cycle and thus helps in maintaining the water balance in the environment. It is also used in various industrial processes by the scientists and engineers for separating mixtures and manufacturing pure substances.

Forms of Condensation

- **Dew**: Condensation occurs when water vapor in the air comes into contact with a cooler surface, such as grass or car windows, during the night, forming dew droplets.
- **Fog**: Fog is formed when condensation happens near the Earth's surface, resulting in the suspension of water droplets in the air, reducing visibility.
- **Clouds**: Clouds form when moist air rises and cools, causing water vapor to condense around tiny particles in the atmosphere, forming visible cloud formations.
- **Frost**: Frost forms when water vapor in the air directly changes into ice crystals on a cold surface, such as leaves or windows, bypassing the liquid phase.
- **Rain**: Rain occurs when condensed water vapor in clouds coalesces into larger droplets and falls to the ground due to gravity.
- **Steam**: When water vapor condenses upon contact with a colder surface, such as a mirror or a window in a steamy environment, it forms visible water droplets known as steam.

Precipitation

- The falling of water from the sky in various forms is known as precipitation.
- They're all formed by clouds in the troposphere, which are around 8 to 16 kilometres (4 to 11 miles) above the ground.
- Precipitation happens when any form of water particle falls from the upper atmosphere to the earth's surface. Frictional drag and gravity induce

the drop to the ground. When a particle falls from the cloud, it creates a turbulent wake behind it, promoting faster and more frequent drops.

- Depending on the atmospheric conditions, the crystallized ice may reach the ground as ice pellets or snow, or it may melt and convert into raindrops before reaching the earth's surface.
- Rain, snow, sleet, freezing rain, hail, snow grains, and diamond dust are all examples of diverse types of precipitation. They are several types of water that fall from the frozen clouds in the sky.

Process of Precipitation

- Precipitation occurs when water vapor in the air mass condenses.
- Rising air masses, carrying a significant amount of water vapor, become saturated due to adiabatic cooling.
- Cloud formation is initiated by the condensation of water vapor, with each cloud containing both updrafts and downdrafts.
- The strength of the updraft influences the growth and height of the cloud; stronger updrafts result in taller clouds. As liquid water accumulates, the updraft weakens and the downdraft strengthens, leading to precipitation.
- Although all clouds contain water, not all produce precipitation. Sometimes, precipitation evaporates before reaching the earth's surface. Precipitation occurs when cloud droplets or ice crystals grow large enough to overcome atmospheric updrafts, indicating specific processes within rain-producing clouds.

Forms of Precipitation

The different forms of precipitation are as follows

Rain

- Rain is a form of precipitation that is in the form of water drops of a size that is larger than 0.5mm. The maximum raindrop size is about 6mm. Drops of larger size break up into smaller drops as it falls down on the earth's surface.
- Rainfall is the predominant form of precipitation and therefore, the term precipitation is used synonymously with rainfall. The magnitude of the rainfall shows high temporal and spatial variation. This variation causes the occurrence of hydrologic extremes like floods and droughts.

Snow

Snow consists of ice crystals in a flaky form, having an average density of 0.1g/cc. It is also an important form of precipitation that usually forms in colder climates and higher altitudes.

Drizzle

Drizzle is a fine sprinkle of tiny water droplets that have a size less than 0.5mm and an intensity greater than 1mm/h. The tiny drops that form a drizzle appear floating in the air.

Glaze or Freezing Rain

The glaze is formed when rain or drizzle comes in direct contact with the cold ground at around 0 degrees celsius. This water drops freeze to form an ice coating known as glaze.

Sleet

Sleet is frozen raindrops that are formed when rainfall passes through the air in the atmosphere at subfreezing temperatures.

Hail

Hail is a kind of showery precipitation in the form of pellets or lumps that have a size greater than 8mm. Hail occurs during violent thunderstorms.

Virga

Virga appears as a streak of rain descending from the bottom of a cloud but evaporates before it touches the ground, typically due to encountering dry or warm air as it descends.

Measurement of Precipitation

The precipitation is measured by a hydrological instrument called a rain gauge. A rain gauge is also variously known as a hydrometer, Ombrometer, or Pluviometer. Rain gauges are mainly of two types.

- ❑ Non-recording or ordinary rain gauges.
- ❑ Recording type or automatic rain gauges

Nonrecording or Original Rain Gages

The non-recording gauge (original gauge) extensively used in the world is the Symon's gauge. Non-recording gauges don't record the rain but only collect the rain. Once the rain is collected, then it is measured by using a graduated cylinder.

The volume of rain measured in the measuring cylinder directly represents the rainfall volume in cm of water depth i.e. the volume of water collected in cm^3 divided by the area of the aperture (hole i.e. opening that admits light) of the gauge in cm^2 is equal to the depth of rainfall in cm.

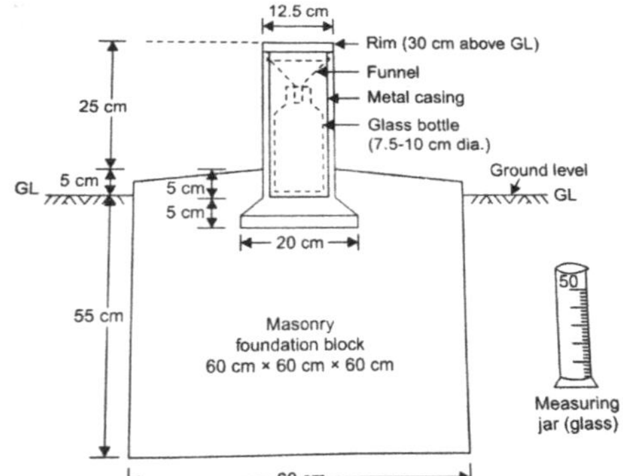

Symon's rain gauge

The receiver (connecting bottle) has a capacity of 175 mm of rain. In a region of heavy rainfall, rain gauge with receivers of 375 mm or 1000 mm capacity to measure rainfall depths may be used.

Nonrecording rain gauges give the amount of rainfall only. This type of rain gauge cannot provide information regarding when exactly the rain commenced, the rain ended, what is the intensity of rainfall, and how much intensity of rainfall varies within the duration of the storm. In order to record the beginning and end of the rain and to measure the intensity of rainfall, a continuous recording of rainfall with time is required, for this purpose, a recording rain gauges is required.

Recording Type or Automatic Rain Gauges

- In the recording gauge, it plots the rainfall against time. From that plot of data, we can easily extract information about the intensity and duration of rainfall, on the basis of which we may make a hydrological analysis of storms.
- Recording-type rain gauges are those rain gauges, which can give for permanent automatic rainfall records without any bottle reading.
- In this type of rain gauge, no man is required to measure or read the amount of rainfall from the rain gauge.
- The record was started and gets recording was recorded automatically on graph paper.
- The gauge, focuses on a record of humidity range versus (VS) time, in the form of a graph, which is known as the mass curve of rainfall.

The following are some of the commonly used recording rain gauges

- Tipping (or tilting) bucket type
- Siphon float type
- Weighing bucket type

Tipping Tilting Bucket-Type Rain Gauges

Collection Funnel: Rainwater is collected through a funnel into a small bucket.

Tipping Mechanism: The bucket is balanced on a pivot. As rainwater accumulates in the bucket, it becomes heavier. Once a specific volume of water (typically 0.2 mm or 0.01 inches) is collected, the bucket tips over due to the imbalance in weight.

Recording Mechanism: When the bucket tips, it triggers a counting mechanism. This mechanism can vary but often involves a reed switch or an optical sensor. These sensors detect the tipping motion and record it as a "tip" event.

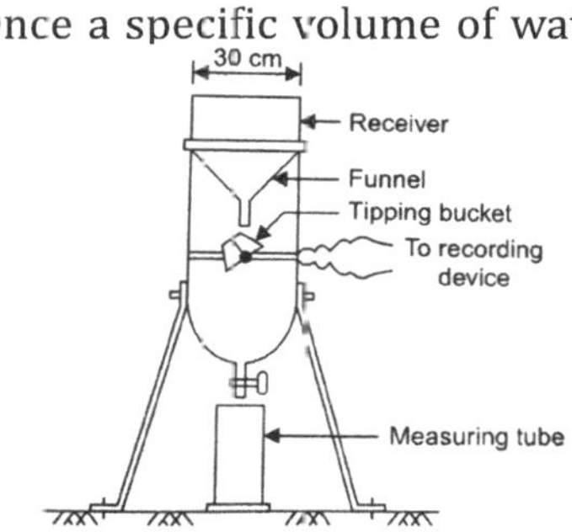

Tipping Bucket Type Rain Gauge

Measurement: Each tip of the bucket corresponds to a known volume of rainfall. By counting the number of tips over a given period, the total rainfall for that period can be calculated.

Merits

- High accuracy in measuring rainfall.
- Real-time monitoring capability.
- Automation reduces the need for manual measurements.
- Standardized measurements for comparison.
- Durability when properly maintained.

Demerits

- Susceptibility to blockage from debris.
- Requires regular maintenance.
- Prone to freezing in cold conditions.
- Limited accuracy for high-intensity rainfall.
- Initial cost and ongoing maintenance expenses

Syphon (Float) Type Automatic Rainfall Recorder

The working principle of the float-type rain gauge resembles that of the tipping bucket-type rain gauge. Rainwater enters the gauge through a funnel, leading into a container where a float is positioned at the bottom. This float is elevated by the collected rainwater.

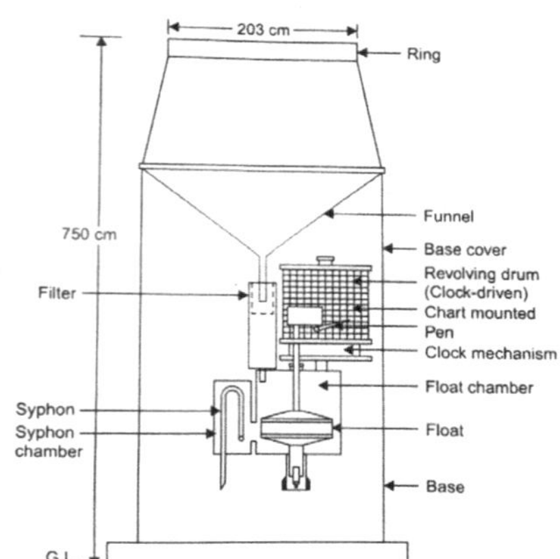

When the float ascends, a pen attached to it via a level system records its elevation on a rotating drum driven by a clock mechanism. Once the float reaches a predetermined maximum level, a siphon arrangement empties the float chamber.

This type of recording rain gauge, also known as a float-type gauge, allows for the collection of various storm-related information. Parameters such as the storm's onset and cessation, its intensity, duration, rainfall distribution, and total precipitation depth can be easily derived from the plotted graph on graph paper.

Merits

- Accurate rainfall measurement.
- Continuous monitoring without manual intervention.

- Reliable performance with minimal maintenance.
- Versatility for various environmental conditions.
- Efficient data collection for analysis.

Demerits

- Higher initial cost.
- Maintenance needs periodic checks.
- Susceptibility to environmental factors.
- Limited data resolution compared to advanced technologies.
- Interpretation may require specialized knowledge.

Waiting Bucket Type Rain Gauges

This rain gauge is designed to measure both rainfall and snowfall. Rain or snow is collected in a receiver attached to a spring balance. A mechanical lever arm connected to the balance moves a pen on a clock-mounted drum with graph paper. As precipitation accumulates, the weight of the receiver changes, causing the pen to mark a line on the graph paper as time passes. The recorded data shows the accumulation of precipitation over time. The recording interval can be set to 24 hours or 7 days depending on the size of the clock and drum.

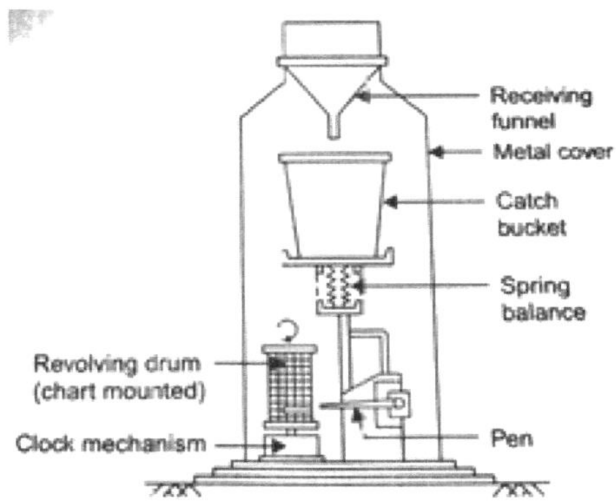

Merits

- Measures both rainfall and snowfall.
- Provides continuous recording without manual effort.
- Offers accurate data on precipitation accumulation.
- Easy interpretation of precipitation patterns.
- Recording intervals can be adjusted.

Demerits

- Limited capacity for heavy rainfall or snowfall.
- Requires periodic maintenance for accuracy.
- Risk of mechanical failure over time.
- Initial setup costs may be high.
- Requires some technical knowledge for setup and maintenance.

Errors in Rainfall Measurement

- Rain gauges typically underestimate actual rainfall.

- Variations in gauge type, height above ground, rainfall intensity, and wind speed can affect measurements.
- Errors may occur during reading or transferring water from the receiver to the measuring jar.

Procedure of Mass Curve Technique:

- Choose at least five reliable base stations in the same climate region.
- Ensure selected base stations have consistent data.
- Calculate annual rainfall averages for each year at the base stations and cumulatively add them.
- Plot a double mass curve graph with base station rainfall on the x-axis and problem station rainfall on the y-axis.

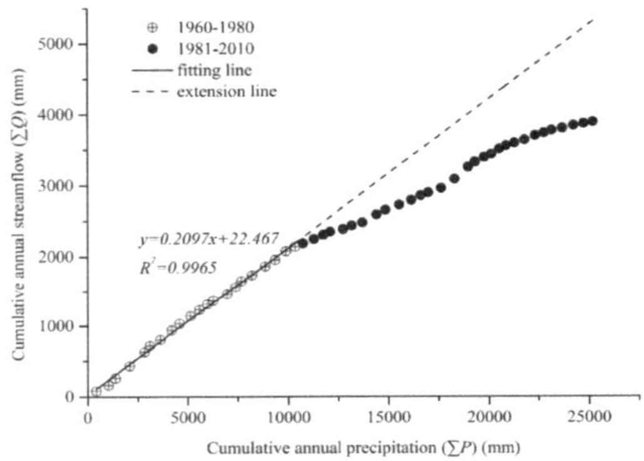

Rainfall in India

- India experiences rainfall through various weather phenomena and seasons.
- The southwest monsoon, occurring from June to September, brings the majority of the country's rainfall.
- It starts in the southern part of India and progresses northward over time.
- Regions like the Western Ghats and northeastern states receive heavy rainfall during the southwest monsoon.
- Northwestern India experiences arid conditions during this time.
- The northeast monsoon, from October to December, affects the southeastern coast, including Tamil Nadu, Andhra Pradesh, and parts of Karnataka.
- Apart from monsoons, pre-monsoon and post-monsoon showers also contribute to India's rainfall.
- Localized weather systems like cyclones can bring heavy rainfall to specific regions at different times.
- Overall, rainfall patterns and amounts vary significantly from region to region and season to season in India.

Chapter-5

Clouds and Thunderstorms

Clouds are formed due to the condensation of water vapour in the atmosphere. They are visible aggregations of tiny water droplets or ice crystals suspended in the air. The process of condensation occurs when warm, moist air rises, cools, and reaches a point where it can no longer hold the moisture, causing it to condense into these visible cloud formations.

Cloud Formation

- Cloud formation is a captivating process where invisible water vapor transforms into visible water droplets or ice crystals.
- Water vapor and aerosol particles, like salt and dust, are ever-present in the air, regularly colliding.
- As the air cools, water vapor condenses and adheres to these aerosol particles.
- This condensation cycle continues as larger water droplets gradually form around the aerosol particles.
- Clouds start to take shape as more water droplets merge.
- Cloud formation typically occurs when the air becomes saturated, unable to hold additional water vapor.
- This saturation and subsequent condensation can happen in two ways: either by increasing the water content in the air, such as through evaporation, or when the air cools to its dew point, the temperature where condensation begins.

- Warmer air can hold more water vapor, but as it rises and cools, its capacity to retain water vapor diminishes, leading to condensation and cloud formation.
- The altitude where the dew point is reached and clouds form is known as the condensation level.

Factors Leading to Cloud Formation

Moisture Content: The presence of water vapor in the atmosphere is essential for cloud formation. This moisture can come from evaporation of surface water bodies, transpiration from plants, or other sources.

Temperature: Cooling of air is necessary for condensation to occur. As air rises and expands, it cools adiabatically, reaching its dew point where condensation begins. This can happen due to various mechanisms such as orographic lifting, frontal lifting, or convection.

Pressure: Changes in atmospheric pressure can influence cloud formation. Areas of low pressure often lead to rising air, which can cool and condense to form clouds.

Condensation Nuclei: Tiny particles suspended in the atmosphere, including dust, pollen, and pollutants, serve as nuclei for water vapor condensation. These particles offer surfaces where water molecules can accumulate, aiding in the formation of cloud droplets.

Wind Shear: Differences in wind speed and direction with height can promote cloud formation by enhancing vertical motion in the atmosphere. Wind shear can contribute to the development of organized cloud structures such as cumulus clouds or thunderstorms.

Stability: Atmospheric stability determines whether air parcels will rise or sink. Unstable air, where a parcel of air is warmer and less dense than its surroundings, tends to rise and form clouds. Conversely, stable air resists vertical motion, inhibiting cloud formation.

Topography: Mountains, hills, and other landforms can force air to rise, leading to cloud formation through orographic lifting. As air ascends a mountain slope, it cools and moisture may condense to form clouds.

Amount of Cloudiness

The fraction of the sky covered by clouds is known as sky cover, cloud cover, or cloud amount. The World Meteorological Organization defines the okta as the unit of measurement for clouds. In the past, cloud cover was expressed in oktas. An okta divides the sky into eight parts, with each part representing the proportion of cloud cover in eighths. Consequently, the term "eighths" is also used interchangeably with okta.

Nowadays, cloud cover is typically expressed as a percentage, ranging from 0% (completely clear sky) to 100% (completely overcast sky).

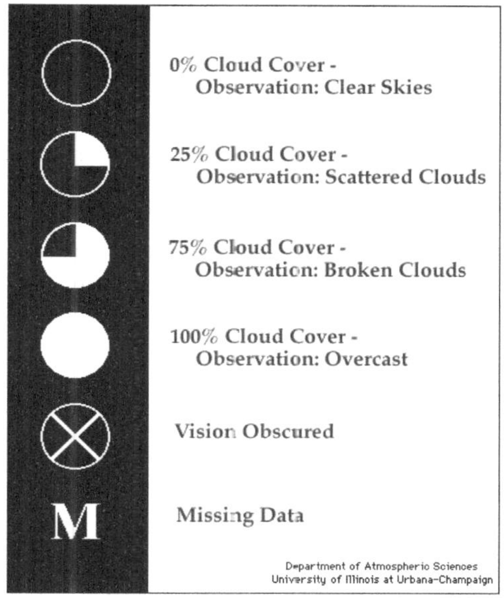

Here's a breakdown of cloud cover percentages and their corresponding descriptions:

- 0%: Clear sky, no clouds visible.
- 25%: Mostly clear, with some scattered clouds.
- 50%: Partly cloudy, with a mix of clouds and clear sky.
- 75%: Mostly cloudy, with significant cloud cover.
- 100%: Overcast, complete cloud cover with no breaks in the clouds.

Ceiling

Ceiling, in the context of aviation and meteorology, refers to the height of the base of the lowest clouds covering more than half of the sky relative to the ground. Here are some types of ceilings related to clouds:

(i) Broken Ceiling

- When more than half of the sky is covered with broken cloud cover (5 to 7 oktas).
- Clouds are fragmented and cover the sky in patches.

(ii) Overcast Ceiling

- When more than half of the sky is covered with overcast cloud cover (8 oktas).
- The sky is completely covered with clouds, and there are no breaks or clear areas.

(iii) Unlimited Ceiling

- Indicates that there is either no cloud cover or that the clouds are at a sufficient height not to impede Visual Flight Rules (VFR) operations.
- Sky appears mostly clear, allowing for unrestricted visibility and flight under VFR conditions.

Classification of Clouds

Clouds are classified based on 2 major factors

- Physical Form
- Height at which Clouds are formed

Classifications of Clouds Based on Their Physical Forms

According to their altitude, stretch, density, and transparency or opaqueness clouds are classified into four types which are given below.

- Cirrus
- Cumulus
- Stratus
- Nimbus

Cirrus clouds

- Cirrus clouds are formed at high altitudes of 8,000 – 12,000m.
- They are detached thin clouds.
- They have a feathery appearance.
- They are always white.

Cumulus

- Cumulus clouds are generally formed at a height of 4,000m – 7,000m.
- They look like cotton wool.
- They exist in patches and can be seen dispersed here and there.
- They have a flat base.

Stratus

- Stratus clouds are horizontal.
- Stratus clouds are stratified or layered clouds covering big portions of the sky.
- These clouds are usually formed due to the mixing of air masses with various temperatures or due to loss of heat.
- The presence of stratus clouds means chilly, overcast day

Nimbus

- Nimbus clouds are usually formed at lower altitudes.
- The colour of Nimbus clouds is usually black or dark grey.
- Nimbus clouds block the sunlight
- These types of clouds usually cause heavy rainfall and thunderstorms.

Classification Based on the Formation

Clouds are classified based on their formation at different heights. The altitude at which certain categories of clouds form varies based on the polar region, tropical region, etc.

Below table gives the classification of clouds and the associated types of clouds

Classification of clouds	Types of clouds
High clouds	Cirrus Cirrostratus Cirrocumulus
Middle clouds	Altostratus Altocumulus
Low clouds	Stratocumulus Nimbostratus
Clouds with extensive vertical development	Cumulus Cumulonimbus

High Altitude Clouds

Cirrus Clouds

- Cirrus clouds are distinct, detached cloud formations characterised by their delicate, white filaments, appearing as patches or narrow bands.
- They often possess a fibrous or silky sheen, with a transparent quality contingent upon the separation of their ice crystals.
- Composed solely of ice crystals, Cirrus clouds typically do not significantly reduce the brightness of the sun when they pass in front of it.
- However, exceptionally thick Cirrus formations may obscure the sun's light and its outline.
- During specific times such as before sunrise or after sunset, Cirrus clouds exhibit striking colours like bright yellow or red.
- These clouds tend to be illuminated earlier than other cloud types, staying visible even after sunset and fading to grey later on.
- When Cirrus clouds are near the horizon, their yellowish hue results from the increased distance and the thickness of air traversed by the rays of light.

Cirrostratus

- Cirrostratus clouds are characterised by their translucent, whitish appearance, often resembling a veil with either a fibrous or smooth texture.
- These extensive sheets of clouds tend to eventually cover the entire sky.
- One way to distinguish a milky veil of fog or a thin stratus cloud from cirrostratus is through the halo phenomena produced by the sun or the moon.

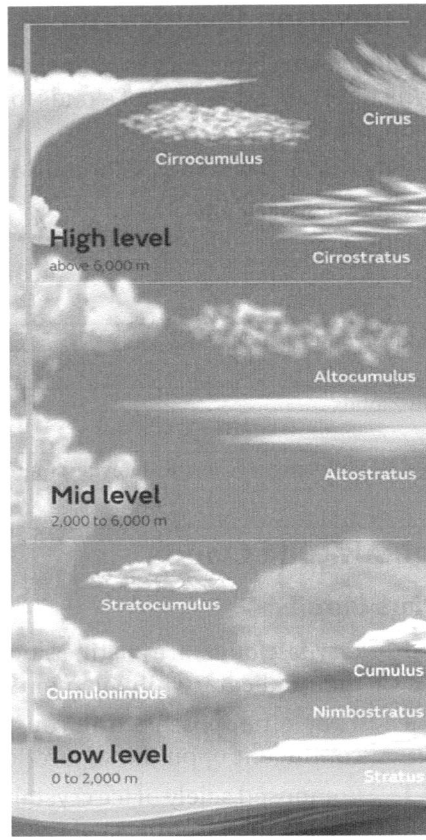

- Cirrostratus clouds commonly generate halos around the sun or moon, distinguishing them from other similar-looking cloud types.
- The presence of a halo effect can often indicate the presence of cirrostratus clouds.

Cirrocumulus

- Cirrocumulus clouds are typically thin, white patches, sheets, or layers in the sky, often not casting shadows.
- They are composed of small elements that can appear in the form of regularly arranged grains or ripples.

Middle Altitude Clouds

Altostratus

- Altostratus clouds often appear as grey or bluish sheets or layers in the sky.
- They can be striated or fibrous and tend to cover the sky partially or completely.
- These clouds are thin enough to allow some sunlight through, giving the impression of viewing the sun through ground glass.
- Unlike cirrostratus clouds, they do not produce a halo effect around the sun or moon, and they also do not permit the shadows of objects on the ground to be visible.
- Altostratus clouds are typically found in the mid-levels of the atmosphere.

Altocumulus

- Altocumulus clouds typically appear as white and/or grey patches, sheets, or layers in the sky.
- They can be made up of laminae (plates), rounded masses, or rolls, and may have fibrous or diffuse edges.
- One distinguishing feature of Altocumulus clouds is their ability to create a colourful optical effect known as a corona.
- When a thin, semi-transparent section of altocumulus clouds passes in front of the sun or moon, a corona, a coloured ring or halo, becomes visible.

- This optical phenomenon features a ring with red on the outer edge and blue toward the centre. Coronas tend to appear within a few degrees of the sun or moon when these types of clouds are present.

Low Level Clouds

Nimbostratus

- Nimbostratus clouds are known as continuous rain clouds, formed by thickening Altostratus clouds.
- These clouds often appear as a dark grey, dense layer diffused by falling rain or snow.
- They are typically thick enough to completely obscure the sun.
- Nimbostratus clouds extend from higher levels into lower altitudes as precipitation continues, gradually merging with low-level clouds.
- They often bring extended periods of steady precipitation, creating overcast and gloomy conditions.

Stratocumulus

- clouds are typically grey or whitish in colour, appearing as patches, sheets, or layers.
- They often exhibit dark tessellations, resembling a honeycomb or mottled appearance, which can be in the form of rounded masses or rolls.
- These clouds usually do not display fibrous elements, except for virga (precipitation that evaporates before reaching the ground).
- Stratocumulus clouds may or may not be merged and are commonly found at lower to middle altitudes.

Vertical Clouds

Cumulus

- These clouds appear as detached, usually dense formations with distinct vertical development, forming rising mounds, domes, or towers, often resembling the shape of a cauliflower, with bulging upper parts.
- The sunlit portions of these clouds often display a brilliant white colour, while their bases appear relatively dark and have a horizontal orientation.

Cumulonimbus

- The cloud being described is a Cumulonimbus cloud, often referred to as a thunderstorm cloud.
- These clouds are heavy and dense, resembling a mountain or a towering structure.
- Their upper portions usually exhibit a smoothed, fibrous, or striated appearance and often display a flattened, anvil-like or plume-like shape.

- Cumulonimbus clouds frequently have very dark bases and are accompanied by low, ragged clouds underneath, which may or may not merge with the base.
- These clouds are known for producing hail and are often associated with tornadoes, creating turbulent and severe weather conditions.

Thunderstorms

- Lightning or electrical storm is known as a thunderstorm. It is characterized by lightning and thunder.
- Thunderstorms produce strong winds, heavy rainfalls, and even produce snowfall and hail. Some of the thunderstorms don't produce any precipitation and, in some cases, thunderstorms produce very little precipitation.

Formations of Thunderstorms

- Cumulonimbus clouds produce thunderstorms and it forms when there is a rapid rise or movement of warm and moist air.
- Cumulonimbus clouds sometimes form at even heights of over 20 kilometres and it is formed when there is the upward movement of warm, moist air. This warm air and moist air cools and condense when it moves upwards resulting in the formation of cumulonimbus clouds.
- Condensation of water vapour into water droplets or ice occur when the rising air reaches its dew point temperature.
- This results in a reduction of pressure locally within the thunderstorm cell.
- Any Precipitation falls a long distance through the clouds on the way to the surface of the Earth. Larger droplets are formed when the smaller droplets falling down collide with each other.
- Downdraft is created by the falling droplets as it pulls cold air with it, and this cold air spreads out at the surface of the Earth. This results in the strong winds commonly produced during thunderstorms.

Stages of Thunderstorms

- Cumulus stage
- Mature stage
- Dissipating stage.

Cumulus Stage

- During the day time, the Earth's surface heats up due to the Sun.
- This leads to updrafts, where warm air ascends upwards. This occurs because the Earth's surface is heated by the Sun, causing the warm air, which is lighter than cooler air, to rise.
- The warm air condenses into cumulus clouds when moist air is present.

- ❑ As long as the warm air beneath, the clouds continues to rise, it will contribute to the ongoing growth of the clouds.

Mature Stage

- ❑ The water in cumulus clouds becomes large and heavy as the cumulus cloud grows in size.
- ❑ Raindrops start to fall through the cloud when the rising air can no longer hold them up.
- ❑ Cool dry air starts to enter the cloud as the raindrops start falling through the clouds.
- ❑ The phenomenon of downdraft takes place. It is a phenomenon that happens when warm air is lighter than the cool air, which makes the cool air to descend in the cloud.
- ❑ Heavy water is pulled downward by the downdraft resulting in rains.
- ❑ Due to updraft, downdraft, and rain; this cloud has become a cumulonimbus cloud.
- ❑ The cumulonimbus cloud has changed into a thunderstorm cell.

Dissipating Stage

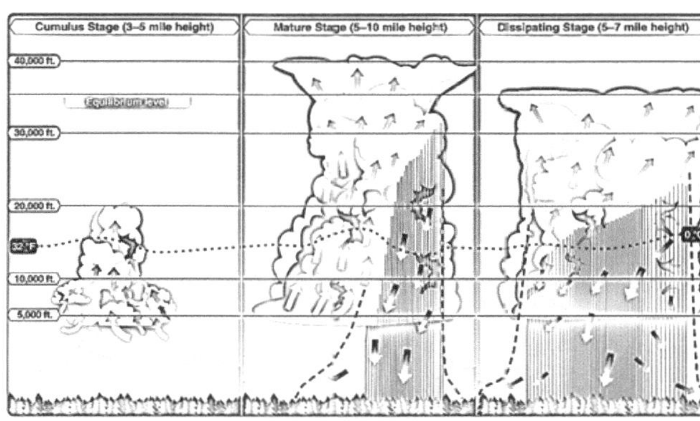

- ❑ Dissipation of thunderstorms happens after a time duration of 30 minutes.
- ❑ This dissipation occurs when the updraft gets dominated by downdrafts in the cloud. Due to this domination of downdraft cloud droplets will cease to form since the warm moist air can no longer rise. As the cloud disappears from top to bottom, the storm will dissipate.
- ❑ For ordinary thunderstorms, the above-mentioned complete process will take approximately 1 hour, whereas for much larger and powerful supercell thunderstorms the above-mentioned dissipation process may take several hours.

Chapter-6

Atmospheric Pressure

Introduction

Atmospheric pressure, or air pressure, is defined as the total weight of a column of air above a unit area at sea level. At sea level, this pressure is equivalent to a 76 cm high mercury column, denoted as 1013.25 mb.

Maximum air pressure is experienced at sea level, where a 1 cm^3 cube weighs approximately 1 kg (1034g). The unit of air pressure is the millibar (mb), with its SI unit being the Pascal.

Air pressure is measured using various instruments, including the normal barometer, Mercury barometer (Fortin barometer), Aneroid barometer, and Micro barograph.

Law of Gases

The gas laws are a group of laws that govern the behaviour of gases by providing relationships between the following:

- The volume occupied by the gas.
- The pressure exerted by a gas on the walls of its container.
- The absolute temperature of the gas.
- The amount of gaseous substance (or) the number of moles of gas.

The gas laws were developed towards the end of the 18th century by numerous scientists (after whom the individual laws are named). The five gas laws are listed below:

Boyle's Law: It provides a relationship between the pressure and the volume of a gas.

$$V \propto 1/P$$

Where V is the volume of the gas, P is the pressure of the gas

Charles's Law: It provides a relationship between the volume occupied by a gas and the absolute temperature.

$$V \propto T$$

Where, V = volume of gas, T = temperature of the gas in Kelvin

Gay-Lussac's Law: Gay-Lussac law gives the relationship between temperature and pressure at constant volume. The law states that at a constant volume, the pressure of the gas is directly proportional to the temperature of a given gas.

$$P \propto T$$

Where, P is the pressure of the gas, and T is the temperature of the gas in Kelvin.

Avogadro's Law: It provides a relationship between the volume occupied by a gas and the amount of gaseous substance.

$$V / n = \text{constant}$$

Where V is the volume of an ideal gas and n represents the number of gas molecules.

Combined Gas Law: The combined gas law, also known as a general gas equation, is obtained by combining three gas laws which include Charle's law, Boyle's Law and Gay-Lussac law. The law shows the relationship between temperature, volume and pressure for a fixed quantity of gas.

$$PV / T = k$$

Ideal gas law: It is the relationship among the four different gas variables such as pressure (P), Volume (V), Temperature (T) and amount of gas (n).

$$PV = nRT$$

Where,

V = volume of gas

T = temperature of the gas

P = pressure of the gas

R = universal gas constant

And n denotes the number of moles

Units of pressure

Pressure is measured by a barometer as the height of mercury column supported by the atmosphere and expressed as a force in millibars (mb) and pounds per square inch (Psi). The unit mb is 1000 dynes per sq.cm.

Pressure Measuring Instruments

Aneroid Barometer

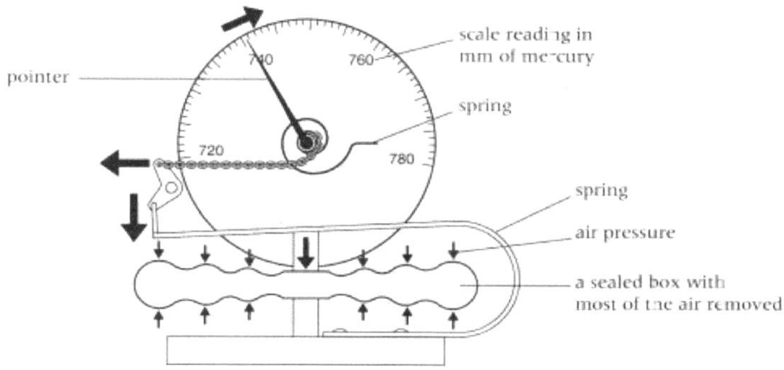

- ❑ A barometer, invented by Italian scientist Evangelista Torricelli in 1643, is a device used to measure air pressure, also known as barometric pressure. As gravity pulls air toward Earth, it exerts pressure on everything it touches. Barometers are instrumental in quantifying this pressure.
- ❑ The horizontal distribution of mean sea level pressure is represented using isobars, connecting locations with equal pressure. When isobars cluster above the normal value in a circular manner, it indicates a high-pressure region.
- ❑ When isobars cluster below the normal value, it indicates a low-pressure region.
- ❑ A ridge occurs when high pressure extends outward into a predominantly low-pressure region, while a trough results from low pressure extending into a region of high pressure.
- ❑ An area of nearly uniform pressure situated between two highs and two lows is referred to as a col.
- ❑ Lines connecting places with equal barometric tendency, where the pressure rises or falls by the same amount in a given period, are termed isobars.

Barograph

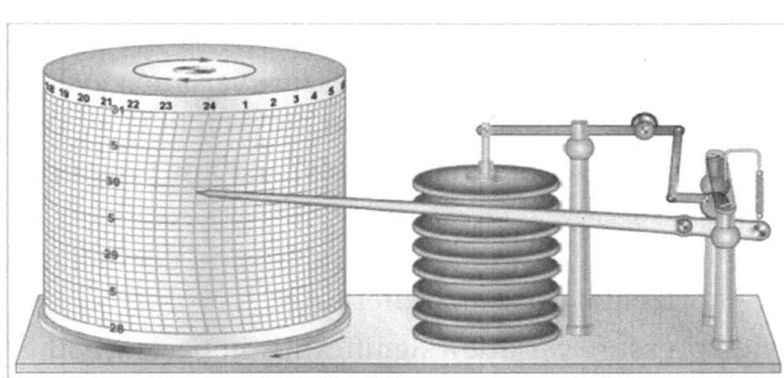

The aneroid barometer may be made to be self-recording so that a trace is drawn on a sheet of paper. The instrument is then terned as a barograph and the record paper is called a barogram.

Isobars

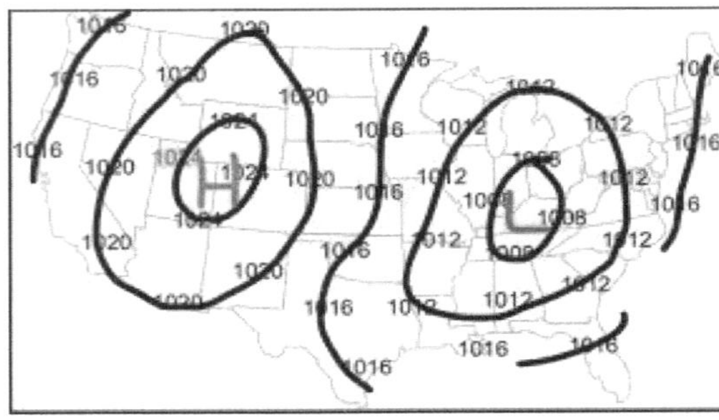

- Isobars are lines linking places with equal pressure. The prefix 'Iso' signifies 'same,' isobar denotes the same pressure. Close-together isobars on a map indicate significant pressure changes over a short distance, while widely spaced isobars suggest pressure changes over a larger distance.
- Wind moves from high-pressure areas to low-pressure areas. The distribution of air pressure is influenced by factors such as altitude, temperature, earth's rotation, water vapor, and air circulation. The rate of pressure change per unit horizontal distance is termed pressure gradient.

Atmospheric Pressure Variation

Vertical Variation

- Atmospheric pressure decreases with increasing altitude.
- Most significant decrease occurs in the troposphere, the layer closest to the Earth's surface.
- Troposphere is where weather phenomena take place.
- Pressure decreases at a slower rate in the stratosphere, mesosphere, thermosphere, and exosphere.
- Understanding vertical pressure variation is crucial for comprehending atmospheric dynamics and phenomena.

Horizontal Variation

- Atmospheric pressure varies horizontally across the earth's surface.
- Factors contributing to horizontal pressure variation include temperature gradients, air density differences, and movement of air masses.
- Regions of high pressure (anticyclones) and low pressure (cyclones) are prominent features.
- Proximity to oceans, latitude, topography, and prevailing wind patterns influence horizontal pressure differences.
- Horizontal pressure variation significantly impacts weather patterns and climate across different geographical areas.

Periodic Variation

- Atmospheric pressure exhibits periodic variations over various time scales.
- Diurnal variations occur daily, driven by solar heating and cooling of the earth's surface.
- Seasonal variations are influenced by changes in solar radiation distribution.
- Weather systems such as high and low-pressure systems, fronts, and storms lead to short-term fluctuations in atmospheric pressure.
- Understanding periodic variations is crucial for accurate weather forecasting and predicting long-term climate trends.

Pressure Gradients

- Atmospheric pressure gradient is the rate of change of atmospheric pressure over a horizontal distance at a given point in time.
- It's typically expressed in millibars per 100 miles.
- The pressure gradient force is driven by differences in barometric pressure between two regions, causing air to flow from high-pressure to low-pressure areas.
- Pressure gradient is steep when pressure changes occur over a short distance and mild when changes occur over longer distances.
- The magnitude of the pressure gradient determines wind speed and direction, influencing weather patterns and conditions.
- Meteorologists use pressure gradient patterns to forecast weather and predict the movement of weather systems, aiding in weather prediction and preparedness efforts.

Pressure Belts

- Atmospheric pressure, influenced by latitudinal variation, temperature, and earth's rotation, exhibits a general pattern across the globe.
- Earth can be divided into regions with broader pressure belts, despite regional anomalies or deviations.
- The pressure belts are not permanent in nature. They oscillate with the apparent movement of the sun. In the northern hemisphere in winter, they move southwards and in the summer northwards. The seven pressure belts are:
  - ☆ Equatorial low-pressure belt
  - ☆ Subtropical high-pressure belt – Northern hemisphere
  - ☆ Subtropical high-pressure belt – Southern hemisphere.
  - ☆ Subpolar low-pressure belt – Northern hemisphere
  - ☆ Subpolar low-pressure belt – Southern hemisphere

- ☆ Polar high-pressure belt – Northern hemisphere
- ☆ Polar high-pressure belt – Southern hemisphere

Four major pressure belts on earth are outlined below:

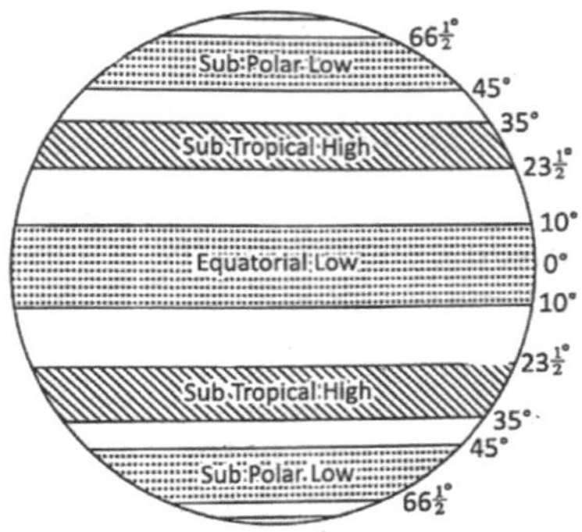

Polar High

Equatorial Low-Pressure Belt

- ❑ Located on either side of the geographical equator.
- ❑ Extends between 5°N and 5°S latitudes, with a potential shift to higher latitudes during respective hemisphere summer.
- ❑ Represents the convergence zone of North-East and South-East trade winds.
- ❑ Also known as doldrums due to frequent calm conditions caused by intense heat leading to low pressure.

Doldrum

- ❑ The equatorial low-pressure belt is located between latitudes 10°N and 10°S.
- ❑ Width fluctuates between 5°N to 5°S and 20°N to 20°S.
- ❑ Convergence of trade winds from the two hemispheres sub-tropical high-pressure belts results in calm air movement, earning it the name Doldrum.

Sub-tropical High-Pressure Belt

- ❑ Extends between latitudes 25° to 35° in both hemispheres.
- ❑ Convergence of winds at higher altitudes leads to the subsidence of air from higher altitudes.
- ❑ Characterized by anti-cyclonic conditions, causing atmospheric stability and aridity due to the descent winds contracting their volume.
- ❑ The zone is named horse latitude due to frequent calm conditions. Dynamic in origin.

Sub-Polar Low-Pressure Belt

- Extends between latitudes 60° to 65° in both hemispheres.
- The sub-polar low-pressure belt is more developed and regular in the Southern hemisphere, while it is fragmented in the Northern hemisphere due to the dominant presence of water in the former.
- Well-defined low-pressure centers or cells are observed over oceans in the Northern Hemisphere, such as in the vicinity of the Aleutian Islands in the Pacific Ocean and between Greenland and Iceland in the Atlantic Ocean.

Polar High-Pressure Belt

- Persistent high pressure at the poles throughout the year due to very low temperatures.
- Both thermal and dynamic factors operate at the poles, with high pressure attributed to the extremely low temperatures.

Shifting of Pressure Belts

- Pressure belts typically shift with the position of the overhead Sun.
- In the Northern Hemisphere, pressure belts move northward in summer and southward in winter.
- Diurnal changes in thermal conditions influence pressure conditions, leading to regional shifts in wind direction.

Chapter-7

Winds

- Wind, or moving air, results from differences in air pressure within the atmosphere. It flows from high-pressure areas to low-pressure areas, and the greater the pressure difference, the faster the air moves.
- The wind is characterized by its direction and speed. The direction is expressed as the direction from which the wind is blowing, such as Easterly winds moving from East to West and Westerly winds moving from West to East.

Factors Affect Wind Speed and Movement

- **The Sun:** Unequal heating of the earth's surface by the Sun plays a crucial role in determining wind speed and movement.
- **Pressure Gradient:** When there is a pressure differential between two regions, air moves from higher pressure to lower pressure due to the pressure gradient force.
- **Coriolis Effect:** Earth's rotation causes objects to deflect to the right in the Northern hemisphere and to the left in the Southern hemisphere. The Coriolis effect influences the direction of movement but not the speed.
- **Friction:** The force that affects wind direction, mainly in the lower troposphere, slowing wind movement and reducing the impact of the Coriolis effect.
- **Centripetal Force:** This force applies when isobars are curved, as in cyclones, pulling air inward along with pressure gradient and Coriolis forces, creating the gradient wind.

Classification of Winds

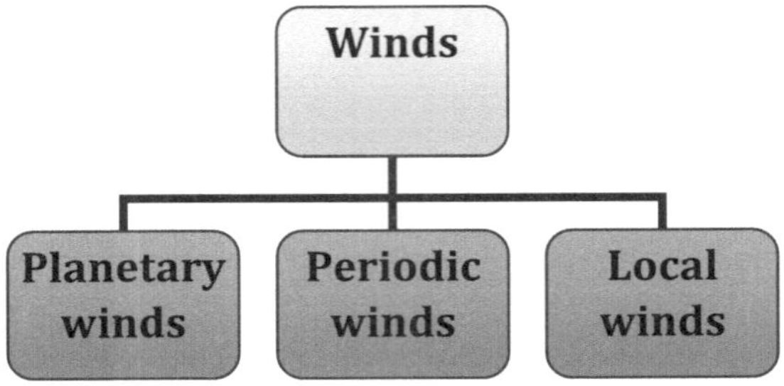

1. Planetary Winds

- The winds that blow consistently throughout the year from one latitude to another in response to variations in air pressure are known as "planetary" or "prevailing" winds.
- They have a global reach and are influenced by the distribution of pressure across the planet. Factors such as the latitudinal differences in atmospheric heating, the formation of pressure belts, the movement of these belts with the apparent path of the sun, the layout of continents and oceans, and the Earth's rotation all play key roles in shaping these winds.
- These winds maintain a constant direction year-round and play a crucial role in transporting energy and sustaining the global heat balance. The planetary winds that circulate across the Earth's surface include

(i) Trade Winds

- Trade winds are extremely steady winds blowing from sub-tropical high-pressure areas (30°N and S) towards the equatorial low-pressure belt.
- Trade winds should have blown from the north to south in the Northern Hemisphere and south to north in the Southern Hemisphere, but they get deflected to the right in the Northern Hemisphere and to the left in the Southern Hemisphere due to Coriolis effect and Ferrel's law. Thus, they blow as north eastern trades in the Northern Hemisphere and south eastern trades in the Southern Hemisphere.
- Trade winds bring maritime conditions to the eastern margins of continents, particularly during the summer months. They converge near the equator, forming the Inter-Tropical Convergence Zone (ITCZ), where these winds lead to heavy rainfall.

(ii) Westerly Winds

- Westerly winds also originate from the subtropical high-pressure belt and move towards the subpolar low-pressure belt and, prevail between 35° to 60 ° latitudes.

- Westerly winds are also permanent, but more intense during winters. They transport warm and moist air toward the pole.
- Westerly winds cause formation of fronts along sub polar low-pressure zone and the transport cyclone toward the western margin.
- The British type of climate or the western European type of climate is produced by the westerly which causes rainfall throughout the year.
- In the southern hemisphere, they are uninterrupted and blow with gale force. They are known as roaring forties along 40°S, furious fifties along 50°S and shrieking sixties along 60°S.

(iii) Polar Easterlies

- The Polar easterlies originate from the Polar high-pressure area and flow towards the temperate low-pressure area. During their journey towards the equator, they are deflected westward, becoming North easterlies in the Northern Hemisphere and South easterlies in the Southern Hemisphere.
- These are extremely cold winds that come from the Tundra and Icecap regions of the poles. The Polar easterlies are more regular in the southern hemisphere in comparison to the northern hemisphere. These polar cold winds converge with the warm easterlies near 60° latitudes and form the Polar front or Mid latitude front. This mid-latitude front becomes the centre of the origin of the temperate cyclones.

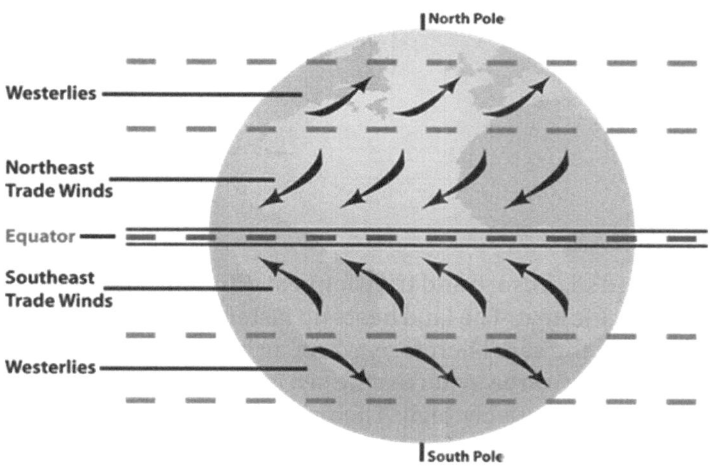

Planetary Winds

2. Periodic Winds/Seasonal Winds

Periodic winds change their direction periodically with the change in season, e.g., Monsoons, Land and Sea Breezes, Mountain and Valley Breezes.

(i) Monsoon Winds

- These are seasonal winds and refer to wind systems that have a pronounced, seasonal reversal of direction.
- According to 'Flohn', monsoon is a seasonal modification of the general planetary wind system. Monsoons cause wet and dry seasons throughout much of the tropics. Monsoons always blow from cold to warm regions. It blows from sea to land during summer & land to sea during winter, due to differential heating of continents & oceans.
- According to Halley's law, in summers, sun shines vertically over the tropic of cancer resulting in high temperature & low pressure in central Asia, while pressure is sufficiently high at the Bay of Bengal & Arabian Sea. This induces air flow from Sea land & induces heavy rainfall in India & neighbouring countries. In winters, sun shines vertically over the Tropic of Capricorn, hence N-W part of India grows colder than Arabian Sea & Bay of Bengal which results in reversal of monsoon in India. Above theory of differential heating was replaced by shifting of the ITCZ for monsoon in India & neighbouring countries.

(ii) Land Breeze and Sea Breeze

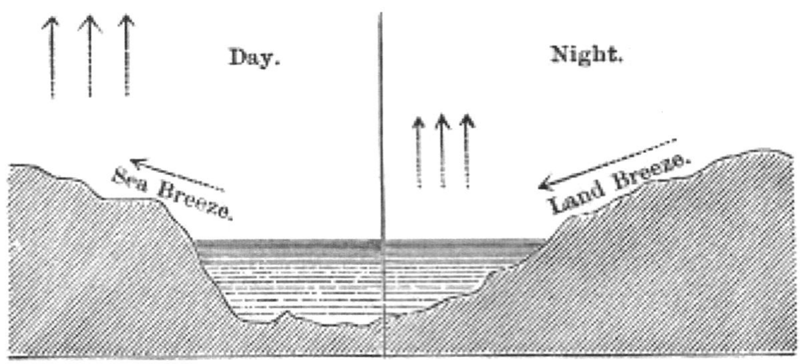

- The land and sea absorb and transfer heat differently due to the difference in their specific heat. The land heats up faster during the day and becomes warmer than the sea. Therefore, warm air rises over the land, giving rise to a low-pressure area, whereas the sea is relatively cool and the pressure over the sea is relatively high. The pressure gradient emerging from sea to land due to the difference in the pressure between sea and land during day time triggers the wind to blow from sea to land. This wind is called sea breeze. During the night, the land loses heat faster and becomes cooler than the sea. Now the pressure gradient is from the land to the sea, which triggers the wind to blow from land to sea. This wind is known as land breeze. Time triggers the wind to blow from sea to land. This wind is called sea breeze.

- During the night, the land loses heat faster and becomes cooler than the sea. Now the pressure gradient is from the land to the sea, which triggers the wind to blow from land to sea. This wind is known as land breeze.

(iii) Valley and Mountain Breeze

- In mountainous regions, the slopes get heated up during the day and air moves upslope and the wind from the valley blows up the valley to fill the resulting gap. This wind is known as the valley breeze.
- The upslope winds driven by warmer surface temperatures on a mountain slope than the surrounding air column are known as anabatic winds.
- During the night, the slopes of mountains get cooled and air over these slopes becomes dense. This dense air descends into the valley which is known as the mountain breeze.
- The cool air, of the high plateaus and ice fields draining into the valley is called katabatic wind.

Valley & Mountain

3. Local Winds

- Local winds are confined to a specific location and are diurnal or monthly in nature.
- They are characterized by diurnal reversal or appear for a short period of time.
- Local winds are produced either due to physiographic features such as mountain & valley or due to distribution of land and sea. They bring diurnal variation in the weather conditions. There are many such winds around the world, some of them cold, some warm, some wet, some dry. There are many hazards associated with these winds as well. Some of the local winds have been discussed below:

(i) Loo

- Loo prevails in the plains of northern India and Pakistan and blows from the west in the months of May and June, usually in the afternoons.

- Sometimes, it is very hot and dry as its temperature invariably ranges between 45°C and 50°C.
- Loo is a harmful wind and may cause sunstroke to people.

(ii) Fohn/Foehn

- Foehn is a hot local wind, which originates during descends of air on the leeward side of a mountain range in the Alps. It is a strong, gusty, dry and warm wind. The temperature of the wind varies between 15°C and 20°C.
- Fohn is a beneficial wind, which helps animal grazing by melting snow and aids the ripening of grapes.

(iii) Chinook

- Chinook is a hot and dry wind just like fohn and prevail in USA and Canada along the eastern slopes of the Rockies. They are more common in winter and early spring.
- Chinook is beneficial to ranchers as it keeps the grasslands clear of snow during much of the winter.

(iv) Mistral

- Mistral is a cold northwest wind that blows from the Alps over France towards the Mediterranean Sea, channeled through the Rhine valley.
- Mistral is very cold and dry blows with high speed. It is a harmful wind and brings snow storms into southern France.

(v) Sirocco

- Sirocco is a hot, dry and dusty winds blowing from the Sahara Desert over central
- Mediterranean and engulfing North Africa and Southern Europe as well.
- Sirocco arises from a warm, dry, tropical air mass that is pulled northward by low- pressure cells moving eastward across the Mediterranean Sea.
- Sirocco is a harmful wind, which causes dusty dry conditions along the northern coast of Africa, storms in the Mediterranean Sea, and cool wet weather in Europe.

(vi) Bise

- Bise is a cold, dry wind which blows from the north-east, north or northwest in the mountainous regions of south eastern France and western Switzerland in winter months.
- The bise is accompanied by heavy cloud.

(vii) Bora

- The bora is a strong, cold and gusty north-easterly wind, which descends to the Adriatic Sea from the Dinaric Alps, the mountains behind the Dalmatian coast (the coast of Croatia).

- The bora begins suddenly and without warning and the cold air typically descends to the coast so rapidly that it has little time to warm up.
- The bora can reach speeds of more than 100 km/h and has been known to overturn vehicles and blow people off their feet.

(viii) Harmattan

- Harmattan is a dry and comparatively cool wind which blows from the east or north- east on the coast of North Africa between Cape Verde and the Gulf of Guinea during the dry season (November to March).
- Harmattan brings dust and sand from the Sahara Desert, often in sufficient quantity to form a thick haze which hinders navigation on rivers.
- Dust and sand are sometimes carried many hundreds of kilometres out to sea by these winds.

(ix) Khamsin

- Khamsin is a hot, dry, dust-laden, southerly wind over Egypt, the Red Sea and eastern parts of the Mediterranean Sea ahead of eastward-moving depressions.
- Khamsin occurs during the period February to June, being most frequent in March and April.
- The Khamsin is a Sirocco wind, whose name comes from the Arabic word for 'fifty', which is approximately the length of time the wind blows for. In the nineteenth Century, the plague was worst in Egypt when the Khamsin was blowing.

(x) Pampero

- Pampero is the name given to severe line squalls in Argentina and Uruguay, particularly in the Rio de la Plata area.
- Pampero are associated with marked cold fronts and are usually accompanied by rain, thunder and lightning, a sharp drop in temperature and a sudden change of wind direction from northerly or north-westerly to southerly or south-westerly.
- Pampero are most likely to occur during the period June to September.

Wind Measuring Instruments

Wind Vane

The wind vane, a time-tested and dependable weather instrument for determining wind direction, typically comprises a long arrow with a tail, allowed to move freely around a vertical post. This arrow consistently aligns with the direction of

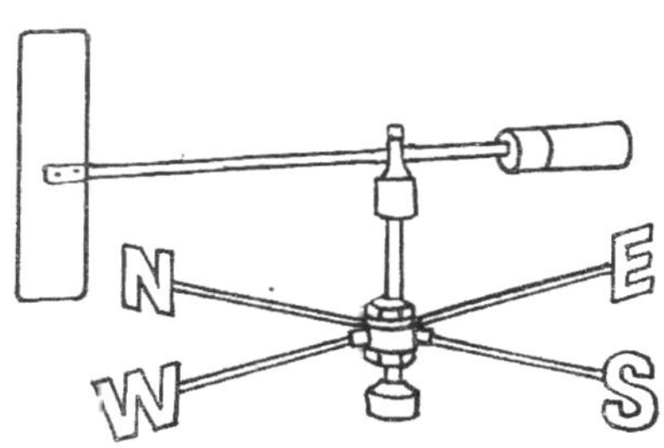

Wind Vane

the wind, providing accurate wind direction information. Wind vanes can be crafted from a variety of materials. At airports, a wind sock, resembling a cone-shaped bag open at both ends, is positioned near the runway. As wind flows through it, the wind sock extends horizontally, aiding pilots in discerning surface wind direction during landing.

Anemometer: The anemometer is the instrument utilized to measure wind speed. Typically, anemometers consist of three or more hemispherical cups mounted on a vertical shaft, known as a cup anemometer. The rotation of the cups is induced by the variance in wind pressure between one side of a cup and the other. However, the wind-measuring devices discussed thus far are considered "ground-based," offering wind speed or direction measurements only at specific fixed locations. Nonetheless, local conditions such as buildings and trees can influence wind patterns. Additionally, wind speed generally escalates rapidly with altitude above ground level. Therefore, it is essential for wind instruments to be positioned in locations where they are exposed to unimpeded airflow well above building rooftops. Unfortunately, in practice, anemometers are often installed at various heights, resulting in inconsistent wind observations

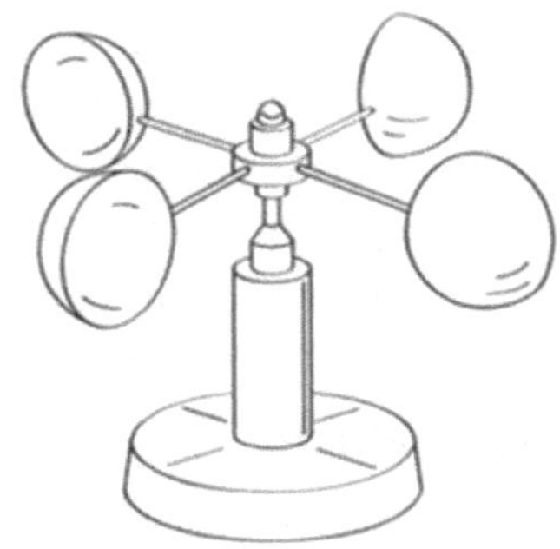

Anemometer

Chapter-8

Tropical Cyclones

Introduction

- Cyclones are rapid inward air circulation around a low-pressure area. The air circulates in an anticlockwise direction in the Northern hemisphere and clockwise in the Southern Hemisphere.
- Cyclones are usually accompanied by violent storms and bad weather.
- The word Cyclone is derived from the Greek word 'Cyclos" meaning the coils of a snake. It was coined by Henry Peddington because the tropical storms in the Bay of Bengal and the Arabian Sea appear like coiled serpents of the sea.

Classification

There are two types of cyclones:

1. Tropical cyclones
2. Extra Tropical cyclones (also called Temperate cyclones or middle latitude cyclones or Frontal cyclones or Wave Cyclones).

- The World Meteorological Organization uses the term 'Tropical Cyclone' to cover weather systems in which winds exceed 'Gale Force' (minimum of 63 km per hour).
- Tropical cyclones develop in the region between the Tropics of Capricorn and Cancer. They are large-scale weather systems developing over tropical or subtropical waters, coalesce into surface wind circulation.
- Extra tropical cyclones occur in temperate zones and high latitude regions, though they are known to originate in the polar regions.

Anti-cyclones

- An anticyclone is the opposite of a cyclone i.e., it has an outward-spiralling air circulation around a high-pressure centre.
- An anticyclone's winds rotate clockwise in the Northern Hemisphere around a center of high pressure.
- In anticyclones, air comes in from above and sinks to the ground. High pressure centers generally have fair weather.

Table: Pattern of Wind Direction in Cyclones and Anticyclones

Pressure System	Pressure Condition at the Centre	Pattern of Wind Direction	
		Northern Hemisphere	Southern Hemisphere
Cyclone	Low	Anticlockwise	Clockwise
Anticyclone	High	Clockwise	Anticlockwise

Tropical Cyclones

- Tropical cyclones are violent storms that originate over oceans in tropical areas and move over to the coastal areas bringing about large-scale destruction caused by violent winds, very heavy rainfall and storm surges.
- Tropical cyclones are one of the most devastating natural calamities in the world.
- Tropical cyclones originate and intensify over warm tropical oceans. The conditions favourable for the formation and intensification of tropical storms are:
  - (i) Large sea surface with temperature higher than 27° C.
  - (ii) Presence of the coriolis force.
  - (iii) Small variations in the vertical wind speed.
  - (iv) A pre-existing weak low- pressure area or low-level-cyclonic circulation. Upper divergence above the sea level system.

Stages of Formation: Tropical Cyclones

The development cycle of tropical cyclone can may be divided into three stages:

Formation and Initial Development Stage

- The formation and initial development of a cyclonic storm depends upon the transfer of water vapour and heat from the warm ocean to the overlying air, primarily by evaporation from the sea surface.
- It encourages formation of massive vertical cumulus clouds due to convection with condensation of rising air above the ocean surface.

Mature Stage

- When a tropical storm intensifies, the air rises in vigorous thunderstorms and tends to spread out horizontally at the tropopause level. Once air spreads out, a positive pressure at high levels is produced, which accelerates the downward motion of air due to convection.
- With the inducement of subsidence, air warms up by compression and a warm 'Eye' (Low pressure centre) is generated. The main physical feature of a mature tropical cyclone in the Indian Ocean is a concentric pattern of highly turbulent giant cumulus thundercloud bands.

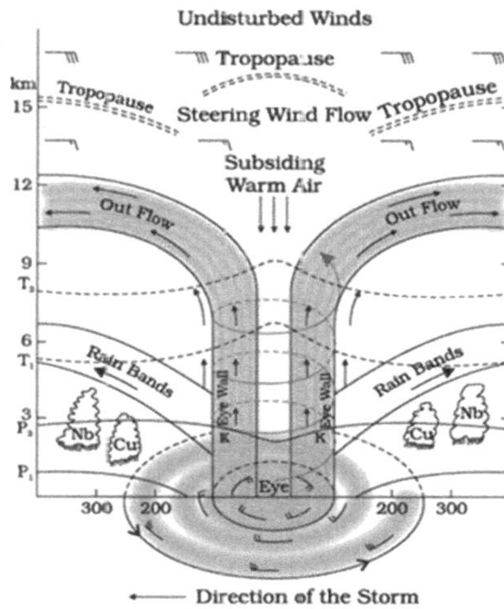

Fig: Vertical of the Tropical Cyclone

Modification and Decay

A tropical cyclone begins to weaken in terms of its central low pressure, internal warmth and extremely high speeds, as soon as its source of warm moist air begins to ebb or is abruptly cut off. This happens after its landfall or when it passes over cold waters.

Nomenclature of Tropical Cyclones

The naming of tropical cyclones is a recent phenomenon. The process of naming cyclones involves several countries in the region and is done under the aegis of the World Meteorological Organization (WMO).

For the Indian Ocean region, a formula for naming cyclones was agreed upon in 2004. Eight countries in the region - Bangladesh, India, Maldives, Myanmar, Oman, Pakistan, Sri Lanka and Thailand - all contributed a set of names which are assigned sequentially whenever a cyclonic storm develops.

Hudhud, Titli, Phethai, Fani, Vayu and Amphan are among the names of cyclones in the Indian Ocean region.

Worldwide Terminology of Tropical Cyclones

They are given many names in different regions of the world – eg. they are known as Typhoons in the China Sea and Pacific Ocean; Hurricanes in the West Indian islands in the Caribbean Sea and Atlantic Ocean; Tornados in the Guinea lands of West Africa and southern USA.; Willy- willies in north-western Australia and Tropical Cyclones in the Indian Ocean.

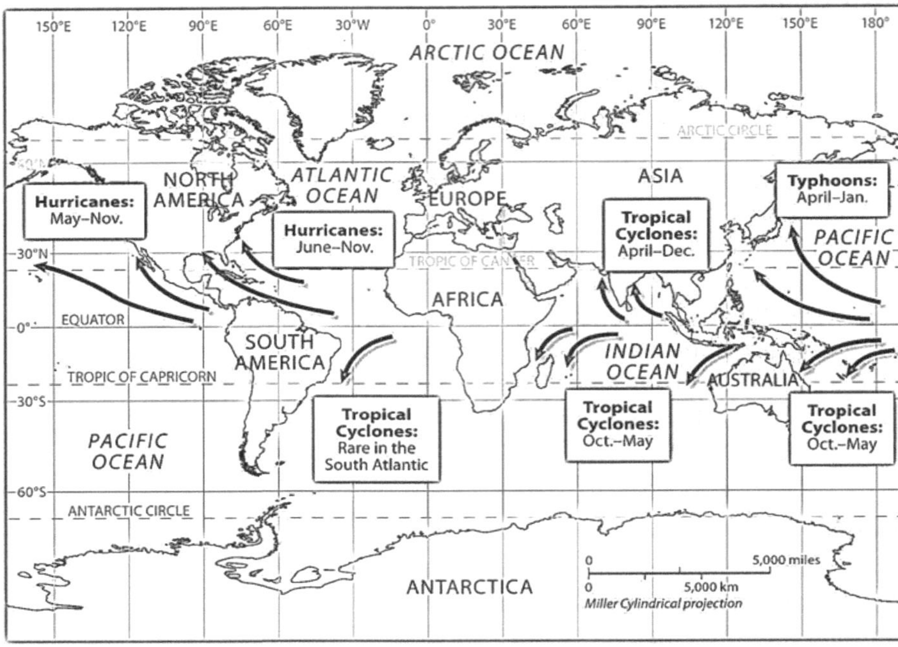

Fig: Distribution of Cyclones in Tropical Regions.

Extratropical Cyclone

- Extratropical cyclones are referred to as mid-latitude depressions, temperate cyclones, frontal depressions and wave cyclones.
- These are active above the mid-latitudinal region between 35° and 65° latitude in both the hemispheres. The direction of movement is from west to east and more pronounced in the winter seasons. In these latitude zones the polar and tropical air masses meet and form fronts.

Formation of Extratropical Cyclones

- The polar front theory describes the origin and evaluation of temperate cyclones. According to this theory, the warm-humid air masses from the tropics meet the dry-cold air masses from the poles and thus a polar front is formed.
- The cold air mass is denser and heavier and due to this reason, warm air mass is pushed up. This interaction of cold and warm air masses creates instability and a low pressure is created at the junction particularly in the center of interactions.
- Reduced pressure creates a void. The surrounding air rushed in to occupy this void and combined with the earth's rotation, a cyclone is formed.

- Extratropical cyclones present a contrast to the more violent cyclones or hurricanes of the tropics, which form in regions of relatively uniform temperatures.

Air mass

- Air Mass is an extremely large body of air with relatively constant temperature and moisture content (humidity), at any given altitude.
- It can cover hundreds of thousands of square miles of area.
- It may have only a little horizontal variation in temperature and moisture throughout the air mass.
- When an air mass remains over a homogenous area for a sufficiently longer time, it acquires the characteristics of the area. The homogenous regions can be the vast ocean surface or vast plains.

Fronts

- When two different air masses (having distinctly different properties) meet, the boundary zone between them is called a front.

There are four types of fronts:

- **Stationary front:** When the front remains stationary, it is called a stationary front.
- **Cold front:** When the cold air moves towards the warm air mass, its contact zone is called the cold front,
- **Warm front:** If the warm air mass moves towards the cold air mass, the contact zone is a warm front.
- **Occluded front:** If an air mass is fully lifted above the land surface, it is called the occluded front.
- The fronts occur in middle latitudes and are characterized by steep gradient in temperature and pressure. It creates changes in temperature and cause the air to rise to form clouds and cause precipitation.

Strom Divisions

The eye

- The eye of a tropical cyclone is a central area characterized by clear skies, warm temperatures, and low atmospheric pressure.
- Surface pressure at the center can drop significantly, leading to intense winds in the surrounding eyewall.
- Within the eye, winds are light due to weak sinking motion, resulting in warmer temperatures compared to other parts of the storm.
- The absence of clouds in the eye is due to warmer air holding more moisture.

- Conditions in the eye may feel oppressive due to the contrast between the calmness within and the high winds and rain in the eyewall.

The Eyewall

- The eyewall is the most dangerous part of a tropical cyclone, featuring the strongest winds, heaviest rainfall, and towering clouds.
- High winds in the eyewall are caused by rapid pressure changes near the eye, peaking above the surface due to less friction.
- Frictional effects near the surface cause winds to turn inward toward the low-pressure eye, extracting heat and moisture from the sea.
- Upward air movement in the eyewall creates powerful vertical winds, crucial for sustaining the storm's intensity.
- At higher altitudes, air flows outward, forming a broad anticyclonic circulation opposite to surface winds.

Rainbands

- Tropical cyclones have deep convective cells around the eye and secondary bands called rainbands that spiral into the center.
- These rotating bands can cause the storm track to appear to wobble, affecting forecasted landfall positions.
- When a cyclone makes landfall, surface friction increases airflow into the eyewall, leading to heavy rainfall exceeding 250 mm (10 inches) in 24 hours.

- In some cases, storms may stall, resulting in prolonged heavy rainfall, with extreme cases reporting up to 760 mm (30 inches) in a five-day period.

Difference Between Tropical Cyclone and Hurricane

Tropical Cyclone	Hurricane
A tropical cyclone is a rapidly circulating storm that develops over warm tropical oceans characterised by high winds, heavy rain and low atmospheric pressure.	Hurricanes are a type of tropical cyclone that are large swirling storms that originate over warm ocean waters.
They occur in the Southeastern Indian Ocean and the Southwestern Pacific Ocean	They occur in the North Atlantic Ocean and the Northeast Pacific Ocean
Regions most affected by the tropical cyclones are India, North-west Australia, Bangladesh, the Indian Ocean islands, such as Mauritius and Madagascar.	The Caribbean Sea Region is most affected by hurricanes.
Tropical cyclones with surface winds less than 39 mph are known as tropical depressions, and those with maximum sustained winds of 39 mph or higher are called tropical storms.	Hurricane's maximum sustained wind speed reaches more than 74 mph
Its rotation is clockwise in the Southern hemisphere and anticlockwise in the Northern hemisphere	Its rotation is clockwise in the Southern hemisphere and anticlockwise in the Northern hemisphere

Cyclones in India

Tropical Cyclones

- Tropical cyclones originate over the Bay of Bengal, Arabian Sea and the Indian ocean. These tropical cyclones have very high wind velocity and heavy rainfall and hit the Indian Coastal states of Tamil Nadu, Andhra Pradesh, West Bengal, Odisha and Gujarat (These five states are more vulnerable to cyclone disasters than others in India).
- Most of these cyclones are very destructive due to high wind velocity and torrential rain that accompanies it.
- There are three elements associated with cyclones which cause destruction during its occurrence. These are-

(i) Strong Winds/Squall: It damages installations, dwellings, communications systems, trees etc., resulting in loss of life and property.

(ii) Torrential rains and inland flooding: Rain is a serious problem for the people who become shelter less due to the cyclone. Heavy rainfall is usually spread over a wide area and causes large scale soil erosion and weakening of embankments.

(iii) Storm Surge: It is an abnormal rise of sea level near the coast caused by a severe tropical cyclone. Due to storm surge sea water inundates low lying areas of coastal regions drowning human beings and livestock, causes eroding beaches and embankments, destroys vegetation and leads to reduction of soil fertility.

Management of Cyclones

- There are many structural and non-structural measures for effective disaster management of cyclones.
- The structural measures include construction of cyclone shelters, construction of cyclone resistant buildings, road links, culverts, bridges, canals, drains, saline embankments, surface water tanks, communication and power transmission networks etc.
- Non-structural measures like early warning dissemination systems, management of coastal zones, awareness generation and disaster risk management and capacity building of all the stakeholders involved.
- These measures are being adopted and tackled on a State-to-State basis under the National Cyclone Risk Mitigation Project (NCRMP) being implemented through World Bank assistance.

Chapter-9

Weather Forecasting

It is the prediction of the weather through application of the principles of physics, supplemented by a variety of statistical and empirical techniques.

In addition to predictions of atmospheric phenomena themselves, weather forecasting includes predictions of changes on Earth's surface caused by atmospheric conditions - e.g., snow and ice cover, storm tides, and floods.

.**For Agriculture and Fisheries:** Farmers and fishers rely on weather forecasts to adjust their activities and optimize production. By anticipating weather patterns, they can plan planting, harvesting, and fishing schedules to maximize yields and minimize losses.

For military strategy: During wartime, accurate weather forecasting allows military strategists to plan battles effectively, considering expected weather conditions. This maximizes the chances of success by leveraging weather-related advantages.

For disaster preparedness and minimizing losses: Weather forecasts enable individuals and communities to prepare for and mitigate the impacts of natural calamities such as floods, typhoons, and storms. This proactive approach helps minimize loss of life and property damage.

For transportation: Weather forecasts significantly influence transportation, particularly in air and water travel. Airlines use weather data to plan flights, considering factors like wind speed, turbulence, and visibility. Similarly, maritime transportation adjusts routes and schedules to avoid storms and rough seas.

For Tourism Promotion: Weather forecasts guide tourism promotion efforts by highlighting favourable conditions for outdoor activities and sightseeing. Tourists are encouraged to visit destinations with predicted pleasant weather, contributing to local economies.

Weather Forecasting Methods

- Weather forecasters depend on a substantial amount of data to create computer models and simulations, aiding in predicting forthcoming weather changes.
- India Meteorological Department (IMD) uses the INSAT series of satellites hovering in the geosynchronous orbit along with the Real-Time Analysis of Products and Information Dissemination (RAPID), a weather data explorer application, that acts as a gateway and provides quick interactive visualisation along with 4-Dimensional analysis capabilities.
- The forecasters use data generated by the satellites around cloud motion, cloud top temperature, water vapour content that help in rainfall estimation, weather forecasting, and provide the genesis of cyclones and their direction.
- Apart from tracking satellite data, IMD collaborates with ISRO for ground-based observations from the Automatic Weather Stations (AWS), the Global Telecommunication System (GTS) that measure temperature, sunshine, wind direction, speed and humidity.
- Meanwhile, the Agro-meteorological Tower (AGROMET) and Doppler Weather Radar (DWR) systems augment the observations.
- In 2021, IMD adopted a new strategy for issuing monthly and seasonal operational forecasts for the southwest monsoon rainfall by modifying the existing two-stage forecasting strategy.
- The new strategy is based on the existing statistical forecasting system and the newly developed Multi-Model Ensemble (MME)-based forecasting system.
- The MME approach uses the coupled global climate models (CGCMs) from different global climate prediction and research centres, including IMD's Monsoon Mission Climate Forecasting System (MMCFS) model.
- All these technological strides have been possible since the National Monsoon Mission (NMM) was initiated in 2012 with a budgetary allocation of Rs 551 crore and the broad objective to set up a dynamical prediction system for seasonal forecast and to improve the monsoon forecasting skills in the country.

India Meteorological Department

- IMD was established in 1875. It is the National Meteorological Service of the country and the principal government agency in all matters relating to meteorology and allied subjects.

- It works as an agency of the Ministry of Earth Sciences of the Government of India.
- It is headquartered in New Delhi.
- IMD is also one of the six Regional Specialized Meteorological Centres of the World Meteorological Organization.

Roles and Responsibilities

- To take meteorological observations and to provide current and forecast meteorological information for optimum operation of weather-sensitive activities like agriculture, irrigation, shipping, aviation, offshore oil explorations, etc.
- To warn against severe weather phenomena like tropical cyclones, norwesters, dust storms, heavy rains and snow, cold and heat waves, etc., which cause destruction of life and property.
- To provide meteorological statistics required for agriculture, water resource management, industries, oil exploration and other nation-building activities.
- To conduct and promote research in meteorology and allied disciplines.

Weather Forecasting Process

The weather forecasting process involves several key steps:

- **Data Collection**: Meteorologists gather data from various sources such as weather stations, satellites, radar systems, weather balloons, buoys, and aircraft. This data includes information on temperature, humidity, wind speed and direction, air pressure, and precipitation.
- **Data Quality Control**: The collected data undergoes quality control procedures to identify and correct errors or inconsistencies. This ensures that only accurate and reliable data is used in the forecasting process.
- **Data Assimilation**: Observational data is assimilated into numerical weather prediction models to initialize them with the most accurate representation of the current state of the atmosphere.
- **Numerical Weather Prediction (NWP)**: Meteorologists use computer models to simulate the behaviour of the atmosphere based on the initial conditions provided by the assimilated data. These models solve complex mathematical equations that describe atmospheric processes such as fluid dynamics, thermodynamics, and radiative transfer.
- **Model Output Post-Processing**: The raw output from the forecast models undergoes post-processing to improve its accuracy and usability. This may involve statistical techniques, bias correction, and the generation of probabilistic forecasts.
- **Interpretation and Analysis**: Meteorologists interpret the model output and analyse forecast guidance to identify patterns, trends, and potential

weather hazards. They use their expertise to generate forecasts tailored to specific regions and applications.

- **Verification**: After the forecast period has passed, meteorologists verify the accuracy of their predictions by comparing them to observed weather conditions. This helps assess the performance of the forecasting models and identify areas for improvement.
- **Communication**: Weather forecasts and advisories are communicated to the public, government agencies, emergency responders, and various industries through various channels such as weather websites, mobile apps, television, radio, and social media.
- **Continuous Monitoring and Updates**: Meteorologists continuously monitor weather conditions and update forecasts as new data becomes available. This allows for timely adjustments to forecasts and warnings as weather patterns evolve.

By following these steps, meteorologists are able to provide accurate and timely weather forecasts to support various societal needs such as agriculture, transportation, energy, and disaster preparedness

Role of Satellite in Weather Forecasting

Satellite	Data Provided	Role in Fisheries Weather Forecasting
NOAA/NASA satellites	Sea Surface Temperature (SST)	SST data is crucial for identifying favourable fishing areas, as it helps determine locations with optimal water temperatures where fish are likely to be abundant.
	Ocean Currents	Ocean current data aids in predicting the movement of fish populations, allowing fishermen to plan their fishing activities based on the expected direction and speed of currents.
	Weather Conditions	Real-time weather data helps fishermen anticipate changes in weather patterns, enabling them to avoid hazardous conditions at sea and ensure the safety of their fishing operations.
Sentinel satellites	High-Resolution Imagery	High-resolution satellite imagery provides detailed information on oceanographic features such as eddies, upwelling zones, and phytoplankton blooms, which are indicators of productive fishing areas.
	Sea Surface Height (Altimetry)	Altimetry data helps in forecasting wave conditions, allowing fishermen to assess the feasibility and safety of fishing operations by predicting wave height and sea state.

Satellite	Data Provided	Role in Fisheries Weather Forecasting
Himawari series	Real-Time Weather Monitoring	Real-time weather data over the Asia-Pacific region assists in predicting weather patterns and cloud cover, providing valuable information for planning fishing activities.
INSAT series	Weather Monitoring Over Indian Ocean	Monitors weather conditions and cloud cover over the Indian Ocean, providing essential data for forecasting weather conditions and planning fishing activities in the region.
METEOSAT series	Continuous Weather Monitoring	Continuous monitoring of weather conditions over Europe, Africa, and parts of the Middle East supports fisheries weather forecasting efforts by providing up-to-date weather information.

Effects of Climate Change on Fisheries Sector

Impact	Description	Example
Shifts in Species distribution	Alterations in the geographical distribution of marine fish species due to changing ocean temperatures, leading to shifts in fishing grounds and catch composition.	Atlantic mackerel moving northward in search of cooler waters as ocean temperatures increase, resulting in changes to traditional fishing grounds and catch locations.
Changes in Fish Migration Patterns	Disruptions in the timing and routes of fish migrations caused by shifts in ocean currents and temperatures, impacting fishing seasons and migration patterns.	Pacific salmon experiencing delayed upstream migration due to warmer water temperatures, affecting fishing seasons and harvest timing for salmon fisheries.
Decreased Fish Stocks	Declines in the abundance of marine fish populations due to climate-induced stressors such as habitat degradation, overfishing, and changes in prey availability.	North Sea cod populations declining due to a combination of overfishing and warmer water temperatures affecting spawning success, leading to reduced catch quotas for commercial fisheries.
Impacts on Fish Reproduction	Climate change affecting the reproductive success of marine fish species through disruptions in spawning behaviours, larval survival rates, and recruitment success.	Coral reef fish experiencing reduced larval survival rates and recruitment due to ocean acidification and warmer water temperatures, impacting population replenishment and fisheries productivity.
Loss of Critical Habitat	Degradation and loss of critical marine habitats such as coral reefs, mangroves, and seagrass beds, reducing breeding, nursery, and feeding grounds for fish species.	Destruction of mangrove forests along coastlines due to sea level rise and storms, reducing nursery habitat for juvenile fish and contributing to declines in fish populations.

Impact	Description	Example
Ocean Acidification	Increased levels of atmospheric CO_2 leading to ocean acidification, negatively impacting the growth, development, and survival of marine organisms, including fish.	Laboratory experiments showing reduced calcification rates in shellfish species such as oysters and mussels due to increased acidity levels in seawater, impacting shell formation and survival.
Altered trophic Interactions	Changes in species composition and abundance affecting trophic interactions within marine ecosystems, altering food webs and prey availability for commercially valuable fish.	Declines in zooplankton populations due to warming ocean temperatures, affecting the primary food source for small pelagic fish such as sardines and anchovies, which in turn impacts the abundance of predatory fish species.
Increased Frequency of Extreme Events	More frequent and intense extreme weather events such as storms, hurricanes, and heatwaves causing damage to fisheries infrastructure and coastal communities.	Destruction of fishing ports and vessels by powerful hurricanes in the Caribbean, disrupting fishing activities and causing economic losses for coastal communities dependent on fisheries.
Economic Impacts on Fisheries	Economic losses for the fisheries sector due to declines in fish stocks, disruptions in fishing activities, and impacts on fishers' livelihoods and related industries.	Declines in shrimp landings and market prices following habitat loss and pollution in coastal areas, leading to reduced income for shrimp fishers and seafood processing industries reliant on shrimp products.
Vulnerability of Coastal Communities	Small-scale fishing communities, particularly in developing countries, facing increased vulnerability to climate change impacts due to reliance on marine resources.	Fishing villages in Bangladesh experiencing food insecurity and economic hardship due to declines in fish catches attributed to overfishing, habitat degradation, and changes in ocean conditions affecting fish abundance.

Chapter-10

Introduction to Geography: Shape, Size and Structure of the Earth

Introduction

- Earth is the only known planet with life, often referred to as the "blue planet" due to its surface being covered by two-thirds water.
- It is the third planet from the sun and the largest of the four terrestrial planets in the solar system.
- Earth is slightly flattened at the poles, giving it a geoid shape, which means it resembles an earth-like shape.
- Its average distance from the sun is about 149 million kilometres or 93 million miles, equivalent to one astronomical unit (AU).
- Estimated to be around 4.5 billion years old, earth formed at the same time as the rest of the solar system.
- Earth revolves around the sun, completing one orbit every 365.25 days, constituting a single earth year.
- Only 3% of Earth's water is fresh, while 97% is salty, and water covers approximately 71% of its surface, with land covering the remaining 29%.
- The earth's atmosphere comprises six layers: the troposphere, stratosphere, mesosphere, thermosphere, exosphere, and ionosphere.
- Earth has one natural satellite, the Moon, along with several temporary artificial satellites orbiting it.

Size and Shape of the Earth

- Earth is not a perfect circle it is an oblate spheroid, it is like a sphere, but the distance from pole to pole is less than the distance around the equator (middle).
- The shape of the earth is called "GEOID" that is, 'an earth-like shape'.
- Earth is bulged out at the equator and flattered at the poles because of centrifugal force.
- The earth spins at a constant rate but the rate of movement is different the equator is moving fastest and the poles are not moving (ignoring the fact that the earth is orbiting the sun). Because of this movement centrifugal force is pulling matter closer to the equator which structures outwards giving the earth a slightly non-spherical shape.
- Geodesy is the science that studies the shape and size of the earth.
- Earth's circumference and diameter differ because its shape is classified as an oblate spheroid or ellipsoid, instead of a true sphere. This means that instead of being of equal circumference in all areas, the poles are squished (slightly flattened at the North and South Poles), resulting in a bulge at the equator, and thus a larger circumference and diameter there.
- The equatorial bulge at earth's equator is measured at 26.5 miles (42.72 km) and is caused by the planet's rotation and gravity.

Motions of the Earth

- Motion is the action or process of moving or changing place or position.
- The earth is constantly in motion, revolving around the Sun and rotating on its axis. These motions account for many of the phenomena we see as normal occurrences: night and day, changing of the seasons, and different climates in different regions.
- The earth has two movements, rotation, and revolution.
- Rotation- A rotation is a circular movement of an object around a center (or point) of rotation.
- A three-dimensional object rotates always around an imaginary line called a rotation axis.
- If the axis is within the body and passes through its center of mass the body is said to rotate upon it or spin.
- Rotation causes days and nights. It takes 23 hours, 56 minutes, and 4.09 seconds for a sidereal day and an exact 24 hours for a mean solar day.
- Earth rotates on its axis from west to east (counter-clockwise). The speed of earth's rotation is 1,674.4 km/h or 1,040.4 miles per hour at the equator.
- The earth has a 23.45° tilt of the axis.

- While the earth is spinning on its axis, it is revolving around the Sun in a counter-clockwise direction. It takes the Earth one full year to complete one full revolution around the Sun. This path is known as the earth's orbit. The mean distance of the Earth from the sun is about 93 milling miles and the distance varies by 3 million miles, forming a slightly oval path.
- The revolution of the earth around the sun traverses a distance of 595 million miles in 365 days, 6 hours, 9 minutes, and 9.5 seconds. This means a speed of 18 miles a second (or 66,000 miles per hour) while at the same time rotating once each twenty-four hours.

Earth rotates in an elliptical orbit around the Sun

- The orbit of the Earth around the sun is elliptical and not circular. Due to this, the distance between the earth and the sun keeps changing.
- When this distance is minimum, the earth is said to be in perihelion (around January 3), When the distance is the maximum, it is said to be in aphelion (around July 4).

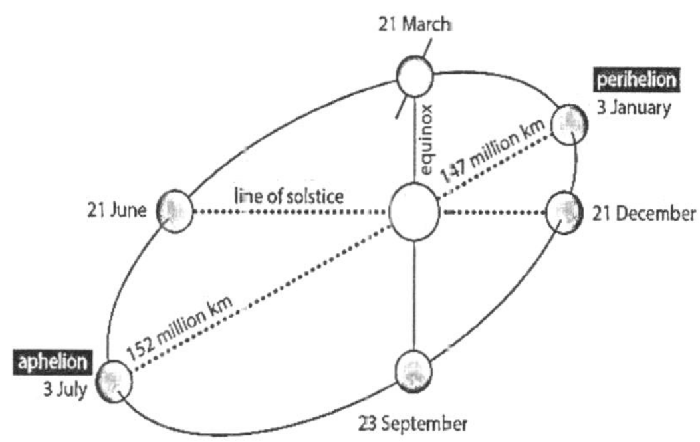

- The Earth's axis points constantly to the same point (the polar star) in the celestial sphere. As a consequence, the latitude on the surface of the earth at which the sun's rays fall vertically keeps changing as the earth moves its orbit around the sun. Due to this, the earth attains four critical positions with reference to the sun.

Equinoxes: Equinox refers to a day with an equal duration of day and night. Two equinoxes in a year are:

- Spring equinox on March 20
- Autumnal equinox on September 22

Solstice: On the other hand, solstice refers to a day with either the longest day or the shortest. The two solstices in a year are:

- Winter solstice on December 22
- Summer solstice on June 21

Difference Between Equinox and Solstice

Equinox	Solstice
Time of the year when the sun is nearest to the equatorial plane giving equal lengths of day and night	Time of the year when the sun is farthest from the equatorial plane resulting in long nights and days
An equinox occurs at the start of the spring and fall	The solstice occurs during the summer and the winter
Happens twice a year	Happens twice a year
Occurs on March 20 (vernal equinox) and on September 22 (autumnal equinox)	Occurs on June 21(Summer Solstice) and on December 22 (Winter Solstice)

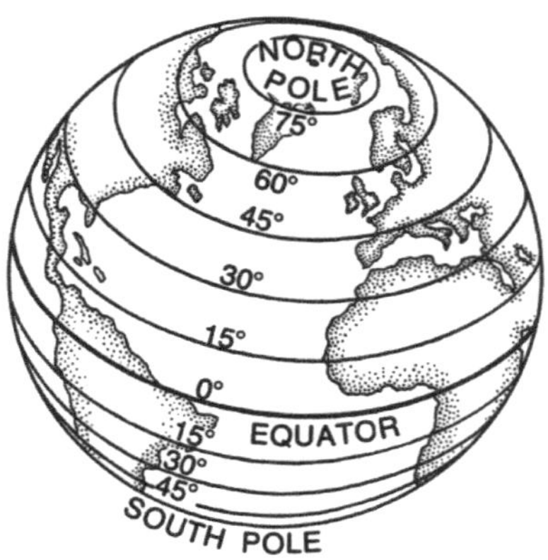

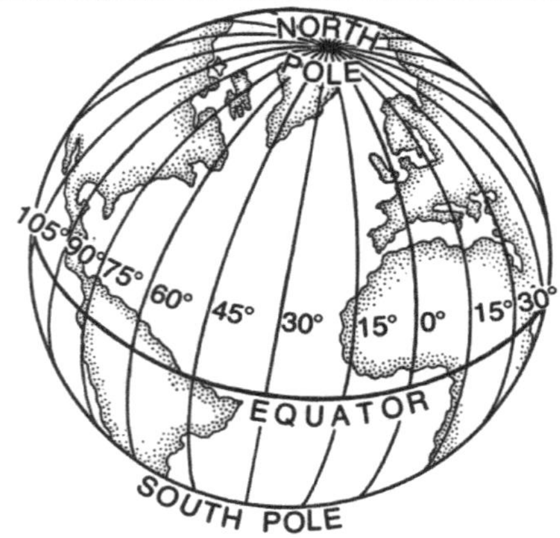

Latitudes and Longitudes

Latitudes and longitudes are imaginary lines used to determine the location of a place on earth.

Latitudes

- The Parallel of latitudes extends from the Equator to 90 degrees North Pole and 90 degrees South Pole.
- If the latitude is drawn at an interval of 1 degree, then in each of the hemispheres there will be 89 latitude lines that will add up to 179 total lines.
- They are mainly the east-west circles that connect all the locations of the Earth.
- The distance between two parallel latitudes is 111km.
- All the latitudes are parallel to the Equator.

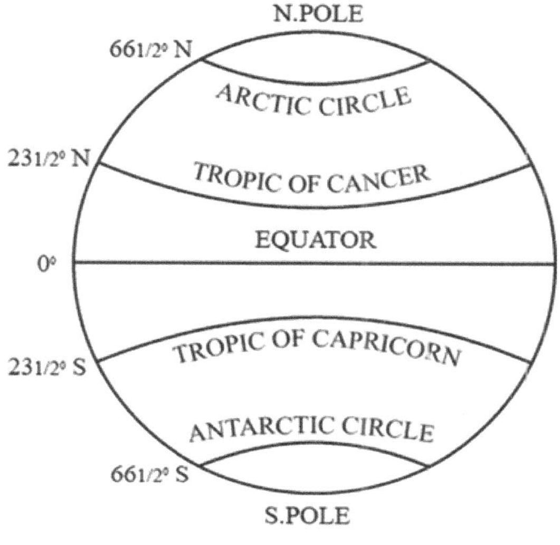

Major parallel of latitudes.

Major Parallel of Latitudes

Equator

- The equator is the imaginary line that divides the earth into two hemispheres.
- The northern hemisphere and the Southern hemisphere.
- It is the longest line of latitude.
- The Equator covers 40075km out of which 78.8% covers the water area while 21.3% covers the surface area.

Arctic Circle

- Out of the five parallel latitudes, Arctic circle is the Northernmost circle which is at 66 and 1/2 degrees north of the equator.
- The position of the arctic is not always fixed.
- As per updates, the arctic is drifting towards the north about 14.5 m (48 ft)/per year.
- The Arctic Circle is 16000km long and covers 4% of the Earth's surface.

Antarctic Circle

- Antarctic circle is the southernmost circle which is currently at 66 and 1/2 degrees south of the equator.
- Like the Arctic circle, the antarctic circle is also shifting southward about 14.5 m (48 ft)/per year.

- The Antarctic Circle is also 16000km long and covers 4% of the Earth's surface towards the south.

Tropic of Cancer

- It is also known as Northern Tropic and the Sun is directly overhead at this position in June.
- The Tropic of Cancer is located at 23 and 1/2 degrees in the Northern Hemisphere.
- The position of the tropic of cancer is fluctuating because of the longitudinal alignment. But the distance between the tropic of cancer and the Antarctic Circle remains the same as they move at the same speed.
- The length of the Tropic of Cancer is 36,788 km.

Tropic of Capricorn

- It is the Southern Tropic which is currently at 23 and 1/2 degree south.
- The length of the Tropic of Capricorn is the same as the Tropic of Cancer.
- The Southern Tropic covers 3% of the total world's population.
- The sun is overhead once a year in December and therefore experiences the maximum heat.

Longitudes

- Longitude is the east-west measurement of the prime meridian.
- There are 180 vertical east longitudes of the Prime Meridian and 180 vertical west longitudes of the Prime Meridian.

Prime Meridian

- The line that passes through the Greenwich, British Royal Observatory is regarded as Prime meridian.
- It is the base Longitude which is 0 degrees from where 180 degrees east and west directions are considered.
- Hence, Prime of Meridian is the base of world time.
- The Prime Meridian divides the Earth into two halves, Eastern and Western hemisphere.

Eastern Hemisphere

- It is the east of the Prime Meridian that covers the country like Africa, Asia, Europe, Australia, and the islands of Oceania.
- The landmass of the Eastern Hemisphere is larger than the western part. Therefore, 80% of the human population survives in the Eastern Hemisphere.
- Eastern Hemisphere is also termed as Oriental Hemisphere.

Western Hemisphere

- It is the west of the Prime Meridian that covers North and South America and some parts of Africa, Europe, Antarctica, and also Asia.
- The center of the Western Hemisphere is in the Pacific Ocean whose nearest land is Genovesa Island.

Longitude and Time

- The best means of measuring time is by the movement of the earth, the moon, and the planets. The sun regularly rises and sets every day.
- When the Prime Meridian of Greenwich has the sun at the highest point in the sky, all the places along this meridian will have mid-day or noon.
- As the earth rotates from west to east, those places east of Greenwich will be ahead of Greenwich time and those to the west will be behind it.
- The earth rotates 360° in about 24 hours, which means 15° an hour or 1° in four minutes. Thus, when it is noon at Greenwich, the time at 15° east of Greenwich will be 15 × 4 = 60 minutes, i.e., 1 hour ahead of Greenwich time, but at 15° west of Greenwich, the time will be behind Greenwich time by one hour.

Great Circle: A great circle is the largest circle that can be drawn on the surface of a sphere. It divides the sphere into two equal halves. In navigation, a great circle route is the shortest distance between two points on the surface of a sphere, such as the Earth. It's often used for long-distance travel like flights between continents or oceanic voyages. While it might look like a straight line on a flat map, it's actually an arc on the curved surface of the Earth.

Rhumb Line: A rhumb line (or loxodrome) is a line that crosses all meridians of longitude at the same angle, i.e., following a constant compass direction. Unlike a great circle route, a rhumb line is a straight line on a Mercator projection map, making it easier for navigators to follow a constant compass bearing. However, it is not the shortest distance between two points on the earth's surface.

Chapter-11

Map Projection

Map projection is a systematic transformation of the latitudes and longitudes of locations on the surface of a sphere or an ellipsoid into locations on a plane. Map projections are necessary for creating maps.

A map projection is the transformation of earth's curved surface (or a portion of) onto a two-dimensional flat surface by means of mathematical equations. During such transformation, the angular geographic coordinates (latitude, longitude) referencing positions on the surface of the Earth are converted to Cartesian coordinates (x, y) representing position of points on a flat map.

Classification of Map Projection

Cylindrical: These projections imagine the earth inside a cylinder, with the equator touching or intersecting the inside surface of the cylinder. They come in three main types: normal, transverse, and oblique. Normal cylindrical projections have their main orientation parallel to the earth's axis, transverse projections are oriented perpendicular to the axis, and oblique projections are neither parallel nor perpendicular.

Conic: In conic projections, the earth's spherical surface is projected onto a cone, which is then unwrapped onto a flat plane. The types of conic projections include normal and secant. Normal conic projections result in either straight or curved meridians that converge at one of the poles, while secant conic projections intersect the sphere along two parallels, reducing distortion.

Azimuthal/Planar: These projections project the earth's surface onto a flat sheet of paper touching the globe. They can be tangent or secant at various points, including the North or South Pole, the equator, or any point in between. Azimuthal projections can have polar, equatorial, or oblique aspects.

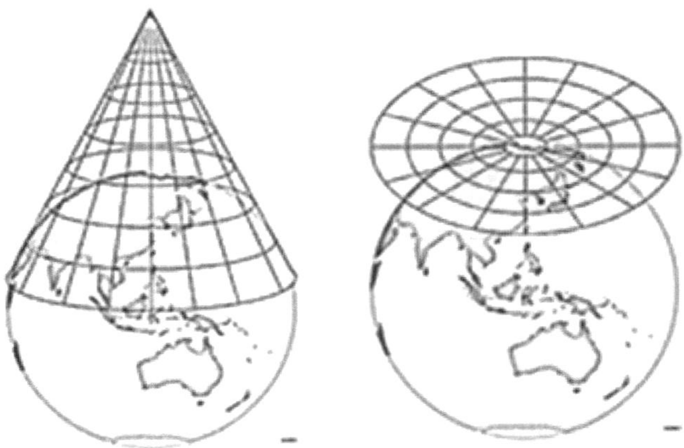

Projection Characteristics

Conformal: Conformal projections maintain local angles, meaning that angles between lines on the map are preserved from the original curved surface. This results in accurate representations of shapes and angles, but areas are distorted, with more significant distortions towards the poles.

Equal Area: Equal area projections aim to preserve the relative sizes of areas on the Earth's surface. While they accurately represent area proportions, other properties such as shape, angle, and scale may be distorted. These projections are useful for maps where showing accurate relative areas is crucial.

Equidistant: Equidistant projections accurately represent distances from the center of projection to any other point on the map. Distances along parallels and meridians are true to scale, making them suitable for applications where measuring distances from a central point is essential.

Commonly Used Projections

Mercator: The Mercator projection is a conformal cylindrical projection that preserves angles, making it suitable for navigation purposes. However, it significantly distorts size as latitude increases, exaggerating areas towards the poles. It's commonly used for mapping equatorial regions but is inappropriate for global mapping due to distortion.

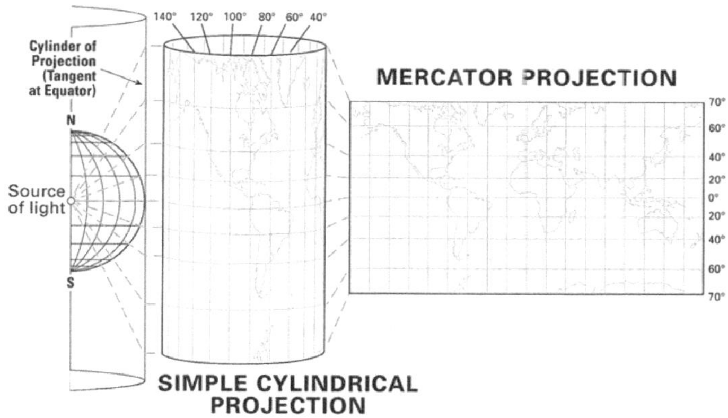

Transverse Mercator: An adaptation of the standard Mercator projection, the Transverse Mercator offers high accuracy in limited zones, making it suitable for topographic maps with significant north-south extents.

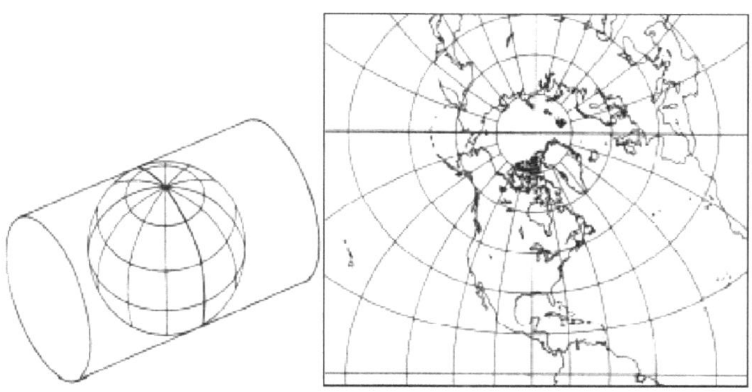

Universal Transverse Mercator (UTM): UTM is an oblique case of the Mercator projection, dividing the world into 60 narrow longitudinal zones. It provides a balance between accuracy and distortion, making it suitable for mapping at various scales. However, it's not suitable for areas beyond certain latitudes due to distortion issues.

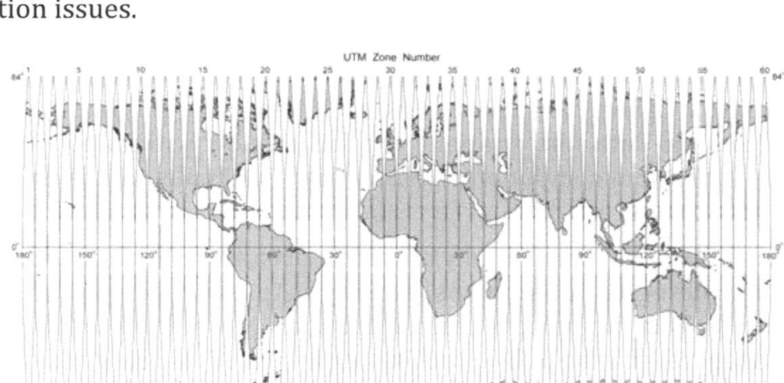

Cartography

Cartography is the discipline of creating, studying, and interpreting maps. It encompasses the science, art, and technology of mapmaking, focusing on representing spatial information accurately and effectively. Cartographers utilize various tools, techniques, and methods to depict geographical features, spatial relationships, and thematic data on maps. These maps serve diverse purposes, including navigation, land-use planning, environmental management, resource exploration, and cultural documentation.

Cartographic Applications in Fisheries

Mapping Fishing Grounds: Cartography plays a crucial role in identifying and mapping fishing grounds, both offshore and inland. By utilizing bathymetric data, satellite imagery, and catch records, cartographers can create maps that depict the spatial distribution of fish populations and the locations of productive fishing areas.

Habitat Mapping: Cartography is used to map important fish habitats such as coral reefs, seagrass beds, mangroves, and estuaries. These habitat maps are essential for understanding the ecological requirements of fish species and for identifying critical areas for conservation and habitat restoration efforts.

Fisheries Management Zones: Cartography is employed to delineate fisheries management zones, which may include areas with specific regulations or restrictions aimed at sustainable fisheries management. These zones can be mapped based on factors such as species distribution, fishing intensity, and habitat sensitivity.

Marine Protected Areas (MPAs): Cartography is used to map marine protected areas, which are designated areas where fishing activities may be restricted or prohibited to conserve marine biodiversity and fish stocks. Maps of MPAs help stakeholders understand their spatial extent and regulations.

Spatio-Temporal Analysis: Cartography enables spatio-temporal analysis of fisheries data, allowing researchers to assess changes in fish populations, fishing effort, and environmental conditions over time and space. By creating time-series maps, cartographers can visualize trends and patterns in fisheries dynamics.

Ecosystem-Based Fisheries Management (EBFM): Cartography supports ecosystem-based fisheries management by integrating biological, ecological, and socioeconomic data into spatially explicit models and maps. These maps help identify ecosystem components, interactions, and vulnerabilities, informing management decisions.

Data Visualization: Cartography is used to visualize fisheries data in a variety of formats, including thematic maps, density maps, and hotspot maps. These visualizations help communicate complex spatial information to policymakers, resource managers, and the public, facilitating informed decision-making.

Landscape

In map projection, a landscape refers to the representation of the Earth's surface on a map. It includes the depiction of various geographical features such as mountains, rivers, valleys, plains, and human-made structures like roads, buildings, and cities. The goal of map projection is to accurately portray these landscape features from the curved surface of the Earth onto a flat map, while accounting factors such as scale, distortion, and projection type.

Application of Landscape in Fisheries

Application	Description
Habitat identification	Using landscape maps to locate and define critical fish habitats such as coral reefs, seagrass beds, and mangroves.
Fishing ground selection	Utilizing landscape maps to identify optimal fishing locations based on underwater topography, currents, and other environmental factors.
Resource allocation	Allocating fishing resources and activities spatially to ensure equitable access and minimize conflicts between different user groups.
Ecosystem-based management	Implementing holistic management approaches that consider the spatial interactions between fish populations, habitats, and environmental factors.
Marine Protected Area Design	Designating protected areas through landscape mapping to conserve biodiversity and enhance fishery sustainability.
Habitat Restoration	Targeting habitat restoration efforts using landscape maps to identify degraded habitats and prioritize restoration activities.

Chapter-12

Disaster Management

Hazard

The UNISDR (United Nations International Strategy for Disaster Reduction) defines a hazard as "a dangerous phenomenon, substance, human activity, or condition that may result in loss of life, injury, health impacts, property damage, loss of livelihoods and services, social or economic disruption, or environmental damage." Hazards encompass both natural and human-caused events that have the potential to escalate into disasters when people are exposed to them. Examples of hazards include droughts resulting from insufficient rainfall, floods triggered by excessive rainfall, or chemical leaks from industrial facilities. When these events occur near human settlements, they pose a threat to human beings, making them hazardous. It's important to note that if such natural or man-made events happen far from human settlements and don't endanger people or property, they are considered mere occurrences rather than hazards. If such events cause widespread negative impacts on humans, they escalate from hazards to disasters.

Types of Hazards

Primarily hazards are of two kinds-natural hazards and human induced hazards. However, disaster may be caused by a combination of both. The UN Inter Governmental Panel on Climate Change (IPCC) has shown that man-made hazards have significantly increased the intensity of disasters. The effects of cyclone, heavy rainfall, earthquake, etc., have been worsened by human activity.

- **Natural Hazard:** Hazard which is the result of natural phenomena is called natural hazard. DesInventar (DesInventar is a conceptual and methodological tool for the construction of databases of loss, damage, or effects caused by emergencies or disasters) has classified natural hazards into five categories:

- **Geophysical:** Geophysical hazards relate to earth phenomena like earthquakes, volcanic activity, tsunami, landslides, land subsidence, etc.
- **Hydrological:** Hydrological hazards are caused by abnormal water cycle like flood, landslide, wave action, etc.
- **Meteorological:** Meteorological hazards are caused due to atmospheric processes like cyclone, storm, lighting, etc.
- **Climatological:** Climatological hazards are caused by long term natural processes like rise in sea level, droughts, glacial lake outburst, etc.
- **Biological:** Biological hazards are caused by bioactive substances (pathogenic micro- organisms). These include epidemics, insect infestation, etc.
- **Human Induced Hazard:** Human induced hazards are caused by human activity. They are also referred as anthropogenic hazards. Some examples of such hazards include, industrial pollution, dam failures, transport or industrial accidents, etc. Most of the man-made disasters are caused by rising population, rapid urbanization and industrialization, environmental degradation, development in high-risk zones, etc. Hazards are also distinguished on the basis of timing:
- **Slow Onset Hazard:** Such hazards develop over a long period of time. Also, onset hazards can be predicted by early warning systems. Examples of such hazards include, droughts, landslides, desertification, cyclones, etc.
- **Rapid Onset Hazard:** Such hazards appear suddenly without early warnings. Examples of such hazard include, fires, flash floods, cloudburst, volcanic eruptions, earthquake, etc.

Disaster Risk

Risk signifies the chances that damage will occur due to exposure to the hazardous situation. Risk is the degree or extent to which human experiences the negative consequences. Accordingly, disaster risk refers to the possibility of suffering damage. Different communities are at different level of risk to suffer the loss due to their vulnerability and coping capacity. Therefore, poorer communities are at greater risk than richer communities who have better capacity to cope with hazards. Thus, the level of risk depends upon: hazardous situation; vulnerability and coping capacity of the given society.

Vulnerability

The term vulnerability implies the possibility of being harmed. In respect to disaster, it signifies the extent of exposure of the people to suffer damage due to hazard. UNISDR explains the vulnerability as "a set of prevailing or consequential conditions arising from various physical, social, economic and environmental factors, which increase the susceptibility of a community to the impact of hazards".

Types of Vulnerability

Physical Vulnerability: This type of vulnerability occurs when nations, organizations, and individuals lack access to basic necessities such as clean water, communication infrastructure, healthcare facilities, security, roads, and electricity.

Political Vulnerability: Political vulnerability arises in environments where democratic processes are absent, preventing citizens from accessing political power and representation. Somalia serves as a prominent example, lacking a stable government structure that allows citizen participation in governance.

Social Vulnerability: Social vulnerability stems from factors like rapid population growth, poverty, ethnic diversity, low education levels, gender inequality, limited access to technology, and breakdown of social structures. This vulnerability may also include religious differences and marginalization.

Economic Vulnerability: Economic vulnerability is determined by factors such as income levels, bargaining power, price levels, inflation, unemployment rates, Gross Domestic Product (GDP), Gross National Product (GNP), and exchange rates.

Environmental Vulnerability: This vulnerability is triggered by land degradation, natural disasters such as typhoons, earthquakes, floods, hurricanes, droughts, storms, deforestation, and wildfires.

Academic Vulnerability: Academic vulnerability arises from a lack of knowledge, expertise, and conceptual understanding needed to solve operational and routine problems at individual, group, organizational, and national levels.

Attitudinal Vulnerability: Attitudinal vulnerability occurs when individuals, organizations, and nations possess a negative attitude towards change and lack initiative and creativity to transform threats into opportunities.

Capacity

Capacity can be defined as "resources, means and strengths which exist in households and communities and which enable them to cope with, withstand, prepare for, prevent, mitigateor quickly recover from a disaster".

Capacities could be:

- Physical capacity
- Socio-economic capacity

Physical Capacity: People whose houses have been destroyed by the cyclone or crops have been destroyed by the flood can salvage things from their homes and from their farms. Some family members have skills, which enable them to find employment if they migrate, either temporarily or permanently.

Socio-economic Capacity: In most of the disasters, people suffer their greatest losses in the physical and material realm. Rich people have the capacity to recover soon because of their wealth. In fact, they are seldom hit by disasters because they live in safe areas and their houses are built with stronger materials.

Disaster

"Disaster is a crisis situation that far exceeds the capabilities."

– Quarentelly, 1985

The term disaster is derived from a French word "Desastre" which implies 'bad/ evil star'. Disasters are debilitating events in which there is widespread loss of life and property in a given society. The social and environmental losses are beyond the capacity of community to cope using its own resources. Therefore, external support is required to overcome its effects. The United Nations Office for Disaster Risk Reduction (UNISDR) defines disaster as:

A serious disruption of the functioning of a community or a society involving widespread human, material, economic or environmental losses and impacts, which exceeds the ability of the affected community or society to cope using its own resources.

The UNISDR definition provides three important components of a disaster:

- There must be a substantial disruption or abnormality in people's lives.
- The community must be incapable of managing the aftermath independently.
- The consequences of the event must be faced by a 'community', i.e., a group of people.

The Disaster Management Act of India defines disaster as

- A catastrophe, mishap, calamity, or significant occurrence in any area resulting from natural or man-made causes, accident, or negligence, which leads to substantial loss of life, human suffering, property damage, destruction, or environmental degradation, and exceeds the coping capacity of the affected community.
- From the above definitions, it can be inferred that a disaster is an event that either causes significant harm to people or disrupts the normal course of life. It is crucial to note that a disaster only occurs when people or public life are affected; without people, there is no disaster. For instance, an earthquake or flood in an uninhabited area cannot be categorized as a disaster.

Causes for Occurrence of Disaster

- **Environmental degradation:** Removal of trees and forest cover from a watershed area have caused, soil erosion, expansion of flood plain area in upper and middle course of rivers and groundwater depletion.
- **Developmental process:** Exploitation of land use, development of infrastructure, rapid urbanization and technological development have caused increasing pressure over the natural resources.

- **Political issues:** War, nuclear power aspirations, fight between countries to become super power and conquering land, sea and skies. These have resulted into wide range of disaster events such as Hiroshima nuclear explosion, Syrian civil war, growing militarisation of oceans and outer space.
- **Industrialization:** This has resulted into warming of earth and frequency of extreme weather events has also increased.

Impacts of Disaster

- Disasters impact individuals both physically (resulting in loss of life, injury, health issues, and disability) and psychologically.
- They lead to significant economic losses due to the destruction of property, human settlements, infrastructure, etc.
- Disasters can alter the natural environment, leading to habitat loss for many plants and animals, and causing ecological stress that may result in biodiversity loss.
- Following natural disasters, food and other essential resources like water often become scarce, resulting in food and water shortages.
- Disasters cause the displacement of people, and those displaced often face numerous challenges in new settlements, exacerbating existing poverty.
- Disasters increase vulnerability levels, thereby amplifying their effects.

Vulnerability Profile of India

- India is vulnerable, in varying degrees, to a large number of disasters. Around 59% of the landmass is prone to earthquakes of moderate to very high intensity.
- About 12% (over 40 million hectares) of its land is prone to floods and river erosion.
- Close to 5,700 kms, out of the 7,517 kms long coastline is prone to cyclones and tsunamis.
- 68% of its cultivable area is vulnerable to droughts; and, the hilly areas are at risk from landslides and avalanches.
- Moreover, India is also vulnerable to chemical, biological, radiological and nuclear (CBRN) emergencies and other man-made disasters.
- Disaster risks in India are further compounded by increasing vulnerabilities related to changing demographics and socio-economic conditions, unplanned urbanization, development within high-risk zones, environmental degradation, climate change, geological hazards, epidemics and pandemics.
- Clearly, all these contribute to a situation where disasters seriously threaten India's economy, its population and sustainable development.

Worst Disasters in India

- **Kerala floods (2018)** resulting from unprecedented rainfall, inflicted widespread devastation in India. Severe flooding and landslides displaced millions, causing hundreds of fatalities and extensive damage to infrastructure and agriculture. The disaster underscored the urgent need for enhanced disaster preparedness and mitigation measures.
- **Kashmir Floods (2014)** affected Srinagar, Bandipur, Rajouri etc. areas of J&K have resulted into death of more than 500 people.
- **Uttarakhand Flash Floods (2013)** affected Govindghat, Kedar Dome, Rudraprayag district of Uttarakhand and resulted into death of more than 5,000 people.
- **The Indian Ocean Tsunami (2004)** affected parts of southern India and Andaman Nicobar Islands, Sri Lanka, Indonesia etc., and resulted in the death of more than 2 lakh people.
- **Gujarat Earthquake (2001)** affected Bhuj, Ahmedabad, Gandhinagar, Kutch, Surat, Surendranagar, Rajkot district, Jamnagar and Jodia districts of Gujarat and resulted in death of more than 20,000 people.
- **Odisha Super Cyclone or Paradip cyclone (1999)** affected the coastal districts of Bhadrak, Kendrapara, Balasore, Jagatsinghpur, Puri, Ganjam etc., and resulted into death of more than 10,000 people.
- **The Great Famine (1876-1878)** affected Madras, Mysore, Hyderabad, and Bombay and resulted into death of around 3 crore people. Even today, it is considered as one of the worst natural calamities in India of all time.
- **Coringa Cyclone (1839)** that affected Coringa district of Andhra Pradesh and **Calcutta Cyclone (1737)** are some other instances of natural calamities faced by the country in the past.
- **The Bengal Famine** in the years **1770** and **1943** affected Bengal, Odisha, Bihar very badly and resulted into death of nearly 1 crore people.
- **Bhopal Gas tragedy (December, 1984)** is one of the worst **chemical disasters** globally that resulted in over 10,000 losing their lives (the actual number remains disputed) and over 5.5 lakh persons affected and suffering from agonizing injuries.
- In recent times, there have been
- cases of railway accidents (Dussehra gathering on the railway tracks crushed by the trains in 2018),
- fire accidents in hospitals due to negligence and non-implementation of existing mandatory fire safety norms,
- collapse of various infrastructure constructs like flyovers, metro tracks and residential buildings due to poor quality of construction, illegal addition of floors and recurring floods.

- Stampede at large public gathering like Kumbh Mela caused by poor people management and lack of adequate infrastructure to monitor and manage large crowd gathering.

Inter-relationship Between Hazard, Vulnerability, Capacity and Disaster Risk

Disaster risk arises from the interaction between hazard and vulnerability. Hazards, on their own, are not inherently harmful. It is only when a society is exposed to a hazard and lacks the capacity to withstand its adverse effects that disaster risk emerges. The following mathematical formula elucidates the relationship between hazard, vulnerability, capacity and disaster risk.

Disaster Risk = (Vulnerability × Hazard)/Capacity

Assessing disaster risk depends on

- Probability of occurrence
- Socio-economic and political conditions
- Intensity and characteristics
- Coping capacity

Types of natural and Manmade Hazards in Fisheries and Aquaculture

1. Cyclone

Definition

A cyclone is a large-scale atmospheric system characterized by a low-pressure center surrounded by rotating winds that spiral inward. Cyclones can form over both land and water, but they are most commonly associated with tropical regions and are known for their strong winds, heavy rainfall, and potential for causing significant damage and disruption to affected areas.

Causes

- Cyclones form when warm, moist air rises over the ocean's surface, creating an area of low pressure.
- The Coriolis effect, caused by the Earth's rotation, causes the air to spin, forming a rotating system known as a cyclone.
- Warm ocean temperatures provide the energy needed for cyclone formation and intensification.

Measurement

- Cyclones are measured based on their wind speed, central pressure, and size.
- Wind speed is measured using instruments such as anemometers, while central pressure is measured using barometers.

- The size of a cyclone can be determined by its diameter or the extent of its wind field.

Adverse Effects

- **Strong Winds:** Cyclones can produce winds exceeding 100 miles per hour, causing extensive damage to buildings, infrastructure, and vegetation.
- **Heavy Rainfall:** Cyclones often bring heavy rainfall, leading to flooding, landslides, and waterlogging in affected areas.
- **Storm Surge:** The low pressure and strong winds associated with cyclones can generate storm surges, resulting in coastal inundation and erosion.
- **Loss of Life and Property:** Cyclones pose significant risks to human life, causing injuries, fatalities, and displacement of populations, as well as damage to homes, businesses, and public infrastructure.

Risk Reduction

- **Early Warning Systems:** Implementing reliable early warning systems to alert communities and authorities about approaching cyclones, allowing time for evacuation and preparedness measures.
- **Infrastructure Resilience:** Constructing resilient infrastructure, including cyclone-resistant buildings, storm shelters, and coastal defences, to withstand cyclone impacts and minimize damage.
- **Coastal Management:** Implementing coastal management strategies such as mangrove restoration, beach nourishment, and dune stabilization to reduce the impacts of storm surges and erosion.
- **Community Preparedness:** Educating and training communities on cyclone preparedness, evacuation procedures, and emergency response measures to enhance resilience and reduce vulnerability.
- **Climate Change Adaptation:** Integrating climate change considerations into cyclone risk management strategies and policies to address future uncertainties and build long-term resilience to changing climate conditions.
- **International Cooperation:** Collaborating with neighbouring countries, regional organizations, and international agencies on transboundary cyclone risk management initiatives, data sharing, and capacity-building efforts to address shared cyclone challenges and promote regional resilience.

2. Floods

Definition: Floods occur when an overflow of water submerges land that is usually dry. They can result from various factors such as heavy rainfall, snowmelt, storm surges, or dam failures, leading to the inundation of low-lying areas and riverbanks.

Causes

- Floods can be caused by:
- **Heavy Rainfall:** Intense or prolonged rain events overwhelm drainage systems and cause rivers to swell.
- **Snow melt:** Melting snow during warmer seasons increases water levels in rivers and streams.
- **Storm Surges:** Tropical storms or hurricanes push seawater inland, causing coastal flooding.
- **Dam or Levee Failures:** Structural failures in dams or levees result in sudden and severe flooding downstream.

Measurement

- Floods are measured by monitoring water levels in rivers, rainfall intensity, and flow rate.
- Hydrological models simulate rainfall-runoff processes to predict flood behaviour.
- Technology such as satellite imagery and tide gauge stations aids in flood monitoring.

Adverse Effects

- **Loss of Life and Property:** Floodwaters can cause drowning, injuries, fatalities, and damage to homes and infrastructure.
- **Disruption of Services:** Flooding disrupts essential services such as water supply, transportation, and communication networks.
- **Environmental Damage:** Floods harm ecosystems, destroy habitats, and contaminate water bodies, affecting biodiversity.
- **Economic Impact:** Floods result in significant economic losses, including damage to agriculture, industry, and property.

Risk Reduction

- **Floodplain Management:** Implementing land use planning and zoning regulations to minimize development in flood-prone areas.
- **Early Warning Systems:** Establishing flood forecasting and warning systems to alert communities ahead of flood events.
- **Infrastructure Development:** Constructing flood control structures such as levees, dams, and stormwater drainage systems.
- **Natural Flood Management:** Restoring wetlands, creating buffer zones, and preserving natural floodplains to absorb floodwaters.
- **Community Preparedness:** Educating and training communities on flood preparedness, evacuation procedures, and emergency response measures.

3. Earthquakes

Definition: Earthquakes are sudden and violent shaking of the ground caused by the movement of tectonic plates beneath the earth's surface. They can range in intensity from minor tremors to devastating seismic events, leading to widespread damage, loss of life, and disruption of communities.

Causes

- **Tectonic Plate Movement:** Earthquakes occur primarily due to the movement, collision, or subduction of tectonic plates along faults, fractures, or plate boundaries in the Earth's crust.
- **Faulting and Fracturing:** Stress accumulation along faults and fractures in the Earth's crust causes sudden release of energy in the form of seismic waves, resulting in ground shaking.
- **Volcanic Activity:** Earthquakes can be triggered by volcanic eruptions, magma movement, or volcanic deformation, particularly in regions with active volcanic systems.
- **Human-Induced Seismicity:** Activities such as mining, reservoir-induced seismicity, hydraulic fracturing (fracking), and geothermal energy extraction can induce earthquakes.

Measurement

- Earthquakes are measured globally using the Moment Magnitude Scale (MMS), which considers the distance moved by a fault and the force applied.
- The USGS reports earthquake magnitudes using MMS, but other scales are used for research and comparison.
- Earthquake intensity is assessed on the Modified Mercalli scale, ranging from I to X (1 to 10) in Roman numerals, commonly used in the United States.
- Seismic waves and ground movement determine earthquake magnitude, typically described on a 10-point scale.

Adverse Effects

- **Structural Damage:** Earthquakes cause structural damage to buildings, infrastructure, roads, bridges, and utilities, leading to collapse, destruction, and loss of property.
- **Casualties and Injuries:** Earthquakes result in casualties, injuries, and fatalities among people caught in collapsing structures, landslides, or other seismic hazards.
- **Tsunamis:** Underwater earthquakes can trigger tsunamis, massive ocean waves that inundate coastal areas, causing flooding, coastal erosion, and additional destruction.

- **Secondary Hazards:** Earthquakes may trigger secondary hazards such as landslides, avalanches, liquefaction, ground subsidence, and aftershocks, exacerbating the impacts and complicating response efforts.
- **Economic Disruption:** Earthquakes disrupt economic activities, commerce, and industry, leading to financial losses, unemployment, and long-term economic recovery challenges.

Risk Reduction

- **Seismic Hazard Assessment:** Conducting seismic hazard assessments, mapping fault lines, identifying high-risk areas, and evaluating building vulnerability to earthquakes.
- **Building Codes and Standards:** Implementing and enforcing seismic building codes, construction standards, and retrofitting requirements to enhance structural resilience and reduce earthquake damage.
- **Early Warning Systems:** Establishing earthquake monitoring networks, sensor networks, and early warning systems to detect seismic activity and issue timely alerts to at-risk communities.
- **Emergency Preparedness:** Developing and implementing emergency response plans, evacuation procedures, and disaster preparedness initiatives to mitigate the impacts of earthquakes and save lives.
- **Public Education and Training:** Educating and training communities, schools, businesses, and emergency responders on earthquake preparedness, safety measures, and response protocols.
- **Land Use Planning:** Incorporating seismic risk considerations into urban planning, infrastructure development, and zoning regulations to minimize exposure and vulnerability to earthquakes.

4. Drought

Definition: Drought is an extended period of abnormally low precipitation, resulting in water shortages and significant impacts on agriculture, ecosystems, and human populations. It is characterized by a deficiency of water supply relative to normal conditions, leading to reduced soil moisture, depleted water sources, and drought-related stresses.

Causes

- **Natural Variability:** Changes in weather patterns, including shifts in atmospheric circulation and precipitation cycles.
- **Below-Average Rainfall:** Prolonged periods of inadequate rainfall or precipitation deficits.
- **High Temperatures:** Elevated temperatures and heatwaves increase evaporation rates and exacerbate water stress.

- **Climate Change:** Altered precipitation patterns and rising temperatures contribute to increased drought frequency and severity.

Measurement

- Droughts are measured using various indicators, including:
- **Precipitation Deficits:** Monitoring rainfall patterns and anomalies relative to historical averages.
- **Soil Moisture:** Assessing soil moisture levels using sensors or satellite imagery.
- **Groundwater Levels:** Monitoring groundwater tables and aquifer levels.
- **Vegetation Health:** Using satellite-based vegetation indices to detect drought-induced stress.

Adverse Effects

- **Agricultural Impacts:** Crop failures, reduced yields, and livestock losses lead to food shortages and economic losses.
- **Water Scarcity:** Shortages for drinking, irrigation, and industry disrupt daily life and economic activities.
- **Ecosystem Degradation:** Droughts damage habitats, increase wildfire risks, and threaten biodiversity.
- **Economic Losses:** Impacts on agriculture, tourism, and other sectors lead to financial strain and unemployment.
- **Social Consequences:** Droughts exacerbate poverty, food insecurity, and health risks in affected communities.

Risk Reduction

- **Drought Monitoring:** Implementing monitoring systems and early warning mechanisms to anticipate drought onset.
- **Water Conservation:** Promoting efficient irrigation, rainwater harvesting, and water reuse/recycling.
- **Drought-Resistant Agriculture:** Developing drought-tolerant crop varieties and sustainable land management practices.
- **Water Management Policies:** Implementing regulations to promote sustainable water use and management.
- **Preparedness Plans:** Formulating drought preparedness and response strategies at local and national levels.
- **Climate Change Adaptation:** Integrating drought risk reduction into climate change adaptation plans and policies.

5. Tsunami

Definition: A tsunami is a series of large ocean waves generated by sudden disturbances on the ocean floor, typically caused by underwater earthquakes, volcanic eruptions, landslides, or meteorite impacts. Tsunamis can travel across entire ocean basins at high speeds and inundate coastal areas with devastating force.

Causes

Tsunamis can be triggered by

- **Underwater Earthquakes:** Sudden movements of tectonic plates along faults beneath the ocean floor.
- **Volcanic Eruptions:** Explosive eruptions or collapses of volcanic islands or underwater volcanoes.
- **Landslides:** Underwater landslides or collapses of coastal cliffs or slopes into the ocean.
- **Meteorite Impacts:** Rare impacts of large meteorites in oceans or seas causing displacement of water.

Measurement

Tsunamis are monitored and measured using

- **Seismic Monitoring:** Detection and analysis of seismic waves generated by underwater earthquakes.
- **Ocean Buoy Systems:** Tsunami detection buoys equipped with pressure sensors to monitor changes in sea level.
- **Tide Gauge Stations:** Measurement of sea level fluctuations using tide gauge stations along coastlines.
- **Satellite Imagery:** Observing and tracking tsunami-related phenomena using satellite-based remote sensing.

Adverse Effects

- **Coastal Inundation:** Tsunamis can inundate coastal areas with rapid influxes of water, causing flooding and erosion.
- **Loss of Life and Property:** Tsunamis pose risks of drowning, injuries, fatalities, and damage to buildings and infrastructure.
- **Environmental Impact:** Tsunamis can disrupt marine ecosystems, damage habitats, and contaminate water bodies.
- **Economic Losses:** Impacts on tourism, fisheries, infrastructure, and coastal development lead to financial losses.
- **Social Disruption:** Tsunamis can cause displacement of populations, social unrest, and psychological stress.

Risk Reduction

- **Early Warning Systems:** Establishing tsunami warning systems to provide timely alerts and evacuation instructions.
- **Infrastructure Resilience:** Constructing coastal defences, seawalls, and tsunami barriers to mitigate impacts.

- **Community Preparedness:** Educating and training communities on tsunami preparedness and evacuation plans.
- **Land Use Planning:** Implementing zoning regulations to restrict development in tsunami-prone areas.
- **International Cooperation:** Collaborating on transboundary tsunami risk management and response initiatives.

6. El-Nino

Definition: El Niño is a climate phenomenon characterized by the abnormal warming of sea surface temperatures in the central and eastern equatorial Pacific Ocean. It occurs irregularly, typically every two to seven years, and can have significant impacts on weather patterns, ocean currents, and global climate variability.

Causes

- **Weakening of Trade Winds:** During El Niño events, the easterly trade winds weaken or reverse direction, allowing warm surface waters to accumulate in the central and eastern Pacific.
- **Oceanic Kelvin Waves:** Changes in atmospheric pressure patterns and oceanic Kelvin waves propagate eastward along the equator, further warming the sea surface temperatures.
- **Southern Oscillation:** El Niño is closely linked to the Southern Oscillation, characterized by variations in atmospheric pressure between the eastern and western Pacific.

Measurement

- El Niño events are measured using indices such as the Southern Oscillation Index (SOI), Equatorial Southern Oscillation Index (EQSOI), and Oceanic Niño Index (ONI).
- Monitoring sea surface temperature anomalies, atmospheric pressure changes, and ocean-atmosphere interactions helps detect and track El Niño events.

Adverse Effects

- **Weather Anomalies:** El Niño disrupts global weather patterns, causing droughts, floods, heatwaves, and other extreme weather events in different regions.
- **Agricultural Impacts:** Crop failures, reduced yields, and food shortages result from altered rainfall patterns and temperature extremes.
- **Ecosystem Disruption:** El Niño affects marine ecosystems, coral reefs, fisheries, and wildlife, leading to habitat degradation and biodiversity loss.
- **Economic Losses:** Impacts on agriculture, fisheries, tourism, and infrastructure result in financial losses and socio-economic challenges.

Risk Reduction

- **Climate Monitoring and Prediction:** Enhancing monitoring and prediction capabilities to anticipate El Niño events and their impacts.
- **Early Warning Systems:** Implementing early warning systems to alert communities and authorities about impending El Niño-related weather extremes.
- **Sustainable Resource Management:** Promoting sustainable land use, water management, and conservation practices to build resilience to El Niño impacts.
- **Disaster Preparedness:** Strengthening disaster preparedness, response, and recovery mechanisms to mitigate the impacts of El Niño-related disasters.
- **International Cooperation:** Collaborating on regional and global initiatives to address climate change, reduce greenhouse gas emissions, and promote climate resilience.

7. Algal Blooms

Definition: Algal blooms are rapid increases in the population density of algae (microscopic photosynthetic organisms) in aquatic ecosystems, leading to the proliferation of algae and the formation of visible surface scums or discolorations in water bodies. These blooms can be composed of various types of algae, including phytoplankton, cyanobacteria (blue-green algae), and other microalgae species.

Causes

- **Nutrient Pollution:** Excessive nutrient inputs, particularly nitrogen and phosphorus from agricultural runoff, sewage discharge, and fertilizer use, can fuel algal growth by providing essential nutrients for photosynthesis and cell proliferation.
- **Warm Water Temperatures:** Elevated water temperatures, combined with ample sunlight and calm conditions, create favourable conditions for algal growth and bloom formation, particularly in stagnant or shallow water bodies.
- **Stagnation and Stratification:** Stagnant or stratified water conditions, where water layers separate based on temperature and density, can inhibit mixing and circulation, allowing algae to accumulate and thrive in surface waters.

Measurement

- Algal blooms can be monitored and measured using various methods, including:
- **Remote Sensing:** Using satellite imagery and remote sensing techniques to detect and map algal bloom extent, density, and distribution in water bodies based on chlorophyll-a concentrations and other optical properties.

- **Water Sampling:** Collecting water samples from affected areas and analysing algal biomass, species composition, cell counts, chlorophyll content, and nutrient concentrations using laboratory techniques such as microscopy, fluorometry, and spectrophotometry.

Adverse Effects

- **Toxicity:** Some algal species can produce toxins, including cyanotoxins, neurotoxins, hepatotoxins, and skin irritants, which pose risks to human health, aquatic organisms, and wildlife through ingestion, inhalation, or dermal contact.
- **Water Quality Degradation:** Algal blooms can degrade water quality by reducing oxygen levels through photosynthetic activity, leading to hypoxia (oxygen depletion), anoxia (complete lack of oxygen), and fish kills in affected water bodies.
- **Ecosystem Imbalance:** Algal blooms can disrupt aquatic ecosystems by outcompeting native species, altering food webs, and depleting resources, leading to declines in biodiversity, habitat degradation, and ecosystem dysfunction.

Risk Reduction

- **Nutrient Management:** Implementing nutrient management strategies, including reducing nutrient inputs from agricultural runoff, sewage discharge, and fertilizer use, to minimize nutrient pollution and mitigate algal bloom risks in water bodies.
- **Wetland Restoration:** Restoring and protecting natural wetlands, riparian buffers, and vegetated shorelines to act as nutrient filters, sediment traps, and buffers against nutrient runoff, promoting water quality improvement and algal bloom prevention.
- **Stormwater Management:** Implementing green infrastructure practices such as rain gardens, permeable pavement, and vegetated swales to capture and treat stormwater runoff, reducing nutrient loading and runoff pollution entering water bodies.

8. Avalanches

Definition: Avalanches are rapid flows of snow, ice, and debris down a mountain slope, often triggered by various factors such as snowfall, temperature changes, or human activities. They can occur in mountainous regions and pose significant risks to people, infrastructure, and the environment.

Causes

- **Snowfall:** Accumulation of snowfall on steep slopes increases the weight and instability of snowpack layers.
- **Temperature Changes:** Rapid temperature fluctuations can weaken snow layers and trigger avalanches.

- **Human Activities:** Activities such as skiing, snowboarding, and snowmobiling can disturb snowpacks and trigger avalanches.
- **Natural Events:** Earthquakes, wind-loading, and rapid melting can destabilize snowpacks and initiate avalanches.

Measurement

- Avalanches are monitored and measured using various methods, including:
- **Snowpack Observations:** Assessing snowpack stability, density, and layering using snow probes, snow pits, and stability tests.
- **Remote Sensing:** Using satellite imagery, aerial surveys, and ground-based sensors to detect and monitor avalanche activity.
- **Avalanche Forecasting:** Meteorological data, terrain analysis, and avalanche bulletins help predict avalanche hazards and inform risk management decisions.

Adverse Effects

- **Loss of Life and Injury:** Avalanches can bury or injure people caught in their path, leading to fatalities and injuries.
- **Property Damage:** Avalanches can damage buildings, infrastructure, roads, and utilities located in avalanche-prone areas.
- **Economic Disruption:** Avalanches disrupt transportation, tourism, and economic activities in affected regions, leading to financial losses.
- **Environmental Impact:** Avalanches can destroy vegetation, alter landscapes, and disturb wildlife habitats, affecting biodiversity and ecosystem functions.

Risk Reduction

- **Avalanche Forecasting:** Providing accurate and timely avalanche forecasts and warnings to inform public safety and risk management efforts.
- **Hazard Zoning:** Implementing land use planning and zoning regulations to restrict development in avalanche-prone areas and protect vulnerable populations.
- **Avalanche Mitigation:** Installing avalanche control measures such as snow fences, barriers, and explosives to reduce avalanche risk and protect infrastructure.
- **Public Education:** Educating and training communities, recreationists, and backcountry users on avalanche awareness, safety practices, and rescue techniques.
- **Search and Rescue:** Establishing coordinated search and rescue teams and protocols to respond to avalanche emergencies and minimize casualties.

9. Pollution

Definition: Pollution refers to the introduction of harmful substances or contaminants into the environment, causing adverse effects on living organisms, ecosystems, and natural resources. Pollution can take various forms, including air pollution, water pollution, soil contamination, and noise pollution.

Causes

- **Industrial Activities:** Emissions from factories, power plants, and manufacturing processes release pollutants such as particulate matter, sulphur dioxide, and heavy metals into the air and water.
- **Transportation:** Vehicle emissions, including exhaust gases and particulate matter from cars, trucks, and airplanes, contribute to air pollution and greenhouse gas emissions.
- **Agricultural Practices:** Pesticides, fertilizers, and livestock waste runoff contaminate soil and water bodies, leading to agricultural pollution.
- **Waste Disposal:** Improper disposal of household, industrial, and hazardous waste contributes to pollution of land, water, and air.
- **Urbanization:** Urban development and population growth result in increased waste generation, resource consumption, and pollution levels in urban areas.

Measurement

- Pollution levels are measured using various methods, including:
- **Air Quality Monitoring:** Monitoring stations measure concentrations of pollutants such as carbon monoxide, nitrogen dioxide, and ozone in the atmosphere.
- **Water Quality Testing:** Sampling and analysis of water samples assess parameters such as pH, dissolved oxygen, and levels of pollutants like heavy metals and organic compounds.
- **Soil Sampling:** Soil tests measure contamination levels of pollutants such as pesticides, heavy metals, and petroleum hydrocarbons in soil samples.
- **Noise Monitoring:** Noise meters measure sound levels and frequencies to assess noise pollution levels in urban and industrial areas.

Adverse Effects

- **Human Health Impacts:** Pollution contributes to respiratory diseases, cardiovascular problems, cancer, and other health issues in humans exposed to pollutants.
- **Environmental Degradation:** Pollution harms ecosystems, degrades habitats, and disrupts biodiversity, leading to species extinction and ecological imbalances.

- **Water Contamination:** Pollutants in water bodies impair water quality, threaten aquatic life, and affect drinking water supplies, fisheries, and recreational activities.
- **Soil Degradation:** Soil pollution reduces soil fertility, contaminates food crops, and poses risks to human health through the food chain.
- **Climate Change:** Pollution, particularly greenhouse gas emissions, contributes to global warming, climate change, and extreme weather events, impacting ecosystems and communities worldwide.

Risk Reduction

- **Pollution Prevention:** Implementing pollution prevention measures such as cleaner production, waste reduction, and recycling to minimize pollutant emissions and waste generation.
- **Regulatory Controls:** Enforcing environmental regulations, standards, and policies to limit pollutant discharges, emissions, and contamination levels.
- **Technology Adoption:** Investing in cleaner technologies, renewable energy sources, and pollution control equipment to reduce emissions and mitigate environmental impacts.
- **Public Awareness:** Educating and raising awareness among individuals, communities, and businesses about the importance of pollution prevention and environmental stewardship.
- **Sustainable Practices:** Promoting sustainable development, resource conservation, and eco-friendly practices to minimize pollution and protect natural resources for future generations.

10. Habitat Destruction

Definition: Habitat destruction refers to the degradation, fragmentation, or loss of natural habitats, including forests, wetlands, grasslands, and coral reefs, due to human activities or environmental changes. It results in the loss of biodiversity, ecosystem services, and wildlife habitats, leading to negative impacts on ecosystems and species populations.

Causes

- **Deforestation:** Clearing of forests for agriculture, logging, urbanization, and infrastructure development results in habitat loss and fragmentation.
- **Urbanization:** Expansion of cities, towns, and infrastructure encroaches on natural habitats, leading to habitat destruction and degradation.
- **Agricultural Expansion:** Conversion of natural ecosystems into farmland, plantations, and grazing lands reduces habitat quality and availability for wildlife.

- **Mining and Extraction:** Extraction of minerals, oil, and natural resources destroys habitats, disrupts ecosystems, and pollutes air and water.
- **Climate Change:** Rising temperatures, changing precipitation patterns, and extreme weather events exacerbate habitat destruction and impact ecosystems.

Measurement

- Habitat destruction is measured using various indicators, including:
- **Land Cover Change:** Satellite imagery and remote sensing techniques track changes in land use and land cover, including deforestation and habitat loss.
- **Habitat Fragmentation:** Spatial analysis measures the degree of habitat fragmentation and connectivity to assess impacts on wildlife populations and biodiversity.
- **Species Decline:** Monitoring populations of indicator species helps identify habitat loss and degradation trends and assess ecosystem health.

Adverse Effects

- **Loss of Biodiversity:** Habitat destruction reduces habitat availability and quality, leading to species loss, extinction, and declines in biodiversity.
- **Ecosystem Disruption:** Habitat destruction disrupts ecological processes, food webs, and nutrient cycles, affecting ecosystem functions and services.
- **Wildlife Decline:** Loss of habitat fragments wildlife populations, restricts movement, and increases susceptibility to predation, disease, and genetic isolation.
- **Soil Erosion:** Habitat destruction increases soil erosion, sedimentation, and nutrient runoff, degrading soil quality and affecting water quality downstream.
- **Climate Change:** Habitat destruction contributes to climate change through the release of carbon stored in vegetation, soil degradation, and loss of carbon sinks.

Risk Reduction

- **Habitat Conservation:** Protecting and preserving natural habitats through conservation measures, protected areas, and habitat restoration initiatives.
- **Land Use Planning:** Implementing land use planning, zoning regulations, and spatial planning to minimize habitat destruction and promote sustainable development.
- **Sustainable Practices:** Adopting sustainable land management practices, including agroforestry, sustainable agriculture, and eco-friendly infrastructure development.

- **Restoration Efforts:** Restoring degraded habitats, reforestation, afforestation, and ecosystem restoration projects to enhance habitat quality and connectivity.
- **Community Engagement:** Involving local communities, stakeholders, and indigenous peoples in habitat conservation and management decisions to promote stewardship and sustainable resource use.

11. Overfishing

Definition: Overfishing refers to the excessive harvesting of fish stocks beyond sustainable levels, resulting in population declines, ecosystem imbalances, and negative impacts on marine biodiversity and fisheries resources.

Causes

- **Increased Demand:** Growing global demand for seafood products, driven by population growth, rising incomes, and changing dietary preferences, leads to intensified fishing pressure.
- **Technological Advances:** Advancements in fishing technology, including larger vessels, mechanized gear, and fish-finding technologies, enable more efficient and widespread fishing operations.
- **Lack of Regulation:** Weak fisheries management, inadequate enforcement of regulations, and illegal, unreported, and unregulated (IUU) fishing practices contribute to overexploitation of fish stocks.
- **Subsidies:** Government subsidies for fuel, vessel construction, and fisheries infrastructure incentivize overfishing and perpetuate unsustainable fishing practices.

Adverse Effects

- **Decline of Fish Stocks:** Overfishing reduces fish populations, disrupts reproductive cycles, and depletes fish stocks, leading to population declines and ecosystem degradation.
- **Ecosystem Imbalance:** Overfishing disrupts marine food webs, alters predator-prey dynamics, and affects ecosystem structure and function, leading to cascading impacts on marine biodiversity.
- **Economic Losses:** Overfishing undermines the long-term sustainability of fisheries, reduces fishery yields, and leads to economic losses for fishermen, fishing communities, and seafood industries.
- **Social Impacts:** Overfishing affects food security, livelihoods, and cultural traditions of coastal communities dependent on fisheries resources for sustenance and income.
- **Illegal Fishing:** Overfishing fuels illegal, unreported, and unregulated (IUU) fishing activities, threatening marine ecosystems, undermining fisheries management efforts, and perpetuating poverty and exploitation in vulnerable communities.

Preventive Measures

- **Sustainable Fisheries Management:** Implementing science-based fisheries management measures, including catch limits, quotas, and fishing licenses, to regulate fishing effort and ensure the long-term sustainability of fish stocks.
- **Monitoring and Enforcement:** Strengthening monitoring, control, and surveillance (MCS) systems to combat illegal fishing, enforce regulations, and deter overexploitation of fish stocks.
- **Marine Protected Areas:** Establishing marine protected areas (MPAs) and no-take zones to conserve critical habitats, protect vulnerable species, and rebuild depleted fish populations.
- **Ecosystem-Based Management:** Adopting ecosystem-based fisheries management (EBFM) approaches that consider the ecological, social, and economic dimensions of fisheries to promote sustainable fishing practices and ecosystem health.
- **Consumer Awareness:** Educating consumers, retailers, and seafood markets about sustainable seafood choices, responsible sourcing, and certification programs to support sustainable fisheries and reduce demand for overexploited species

12. Introduction of exotic species

Definition: The introduction of exotic species refers to the deliberate or accidental introduction of non-native fish species into aquatic ecosystems for purposes such as aquaculture, sport fishing, or biological pest control. These introductions can have significant impacts on native fish populations, aquatic habitats, and ecosystem dynamics.

Causes

- **Aquaculture Practices:** Exotic fish species are often introduced into aquaculture facilities for cultivation and commercial production to meet demand for seafood products.
- **Sport Fishing:** Exotic fish species may be introduced for recreational fishing purposes to enhance angling opportunities and attract tourists.
- **Biological Control:** Non-native fish species may be introduced as biological control agents to manage pest species or control aquatic weeds and algae.
- **Accidental Releases:** Escapes from aquaculture facilities, accidental releases from aquariums, or deliberate releases by hobbyists contribute to the introduction of exotic species.

Adverse Effects

- **Competition and Predation:** Exotic fish species may outcompete native fish for resources such as food, habitat, and breeding sites, leading to declines in native fish populations.

- **Predation Pressure:** Exotic predatory fish may prey on native species, including juveniles and vulnerable life stages, disrupting food webs and ecological balances.
- **Disease Transmission:** Exotic species may introduce new diseases, parasites, or pathogens to native fish populations, leading to disease outbreaks and population declines.
- **Genetic Pollution:** Interbreeding between exotic and native species may result in hybridization, genetic introgression, and loss of genetic diversity in native populations.
- **Habitat Alteration:** Exotic fish species may modify aquatic habitats through foraging, nesting, or spawning behaviours, leading to habitat degradation and ecosystem disruption.

Preventive Measures

- **Risk Assessment:** Conducting risk assessments to evaluate the potential ecological, economic, and social impacts of introducing exotic species into fisheries.
- **Regulation and Legislation:** Enacting regulations and legislation to control the import, transport, and release of exotic species and prevent unauthorized introductions.
- **Biosecurity Protocols:** Implementing biosecurity measures in aquaculture facilities, hatcheries, and fisheries to prevent escapes and accidental releases of exotic species.
- **Monitoring and Surveillance:** Establishing monitoring programs to detect and monitor exotic species populations and assess their spread and impacts on native ecosystems.
- **Public Awareness and Education:** Raising awareness among fishery stakeholders, aquaculturist, and the general public about the risks of introducing exotic species and the importance of responsible management practices.
- **Restoration and Control:** Investing in habitat restoration efforts and control measures to manage exotic species populations and minimize their impacts on native fisheries and ecosystems.

13. Landslides

Definition: Landslides are the movement of a mass of rock, soil, or debris down a slope under the influence of gravity. They can vary in size and speed, ranging from slow-moving slumps to rapid, catastrophic events.

Causes

- **Heavy Rainfall:** Intense or prolonged rainfall can saturate soil and trigger landslides, especially on steep slopes or in areas with poor drainage.

- **Earthquakes:** Seismic activity can destabilize slopes, causing rockfalls, rockslides, or ground failures that result in landslides.
- **Soil Erosion:** Gradual erosion of soil and sediment by water, wind, or human activities weakens slope stability and increases landslide susceptibility.
- **Human Activities:** Deforestation, mining, road construction, and urbanization alter natural landscapes, disturb slopes, and increase landslide risk.

Adverse Effects

- **Loss of Life and Property:** Landslides pose risks to people, buildings, infrastructure, and transportation networks in affected areas, leading to injuries, fatalities, and property damage.
- **Disruption of Services:** Landslides disrupt essential services such as water supply, power, communication, and transportation, causing economic losses and social disruption.
- **Environmental Damage:** Landslides degrade ecosystems, damage vegetation, and alter landscapes, leading to habitat destruction, soil erosion, and sedimentation in water bodies.
- **Economic Impact:** Landslides result in financial losses due to damage to homes, businesses, infrastructure, agriculture, and tourism in affected regions.
- **Geohazard Cascading Effects:** Landslides may trigger secondary hazards such as floods, debris flows, and tsunamis, amplifying their impacts and complicating response efforts.

Risk Reduction

- **Hazard Mapping:** Identifying and mapping landslide-prone areas using geological surveys, remote sensing, and GIS technology to inform land use planning and development regulations.
- **Slope Stabilization:** Implementing slope stabilization measures such as retaining walls, rock bolts, soil nails, and vegetation reinforcement to improve slope stability and reduce landslide risk.
- **Early Warning Systems:** Establishing monitoring networks and early warning systems to detect signs of slope instability and issue timely alerts to at-risk communities.
- **Land Use Planning:** Implementing zoning regulations, building codes, and land management practices to limit development in landslide hazard areas and protect vulnerable populations.
- **Public Education:** Raising awareness among residents, landowners, and stakeholders about landslide hazards, preparedness measures, and evacuation procedures to enhance community resilience.

- **Climate Change Adaptation:** Integrating landslide risk assessment and management into climate change adaptation strategies to address increased precipitation, extreme weather events, and changing environmental conditions.

14. Epidemics

Definition: Epidemics refer to the occurrence of a sudden increase in the number of cases of a disease within a specific population or geographical area, exceeding what is normally expected. Epidemics can range in scale from localized outbreaks to widespread pandemics affecting multiple regions or countries.

Causes

- **Infectious Agents:** Epidemics may be caused by infectious agents such as bacteria, viruses, parasites, or fungi that can spread rapidly among susceptible individuals through direct or indirect transmission.
- **Environmental Factors:** Environmental conditions such as climate, weather, pollution, and habitat changes can influence the transmission dynamics and emergence of epidemic-prone diseases.
- **Population Factors:** Population density, mobility, immunity levels, healthcare access, and behavior patterns can affect disease transmission and susceptibility, contributing to epidemic spread.
- **Globalization and Travel:** Increased international travel, trade, and migration facilitate the spread of infectious diseases across borders, leading to epidemics with global implications.

Adverse Effects

- **Public Health Crisis:** Epidemics strain healthcare systems, overwhelm medical facilities, and challenge public health resources, leading to shortages of medical supplies, personnel, and hospital beds.
- **Morbidity and Mortality:** Epidemics result in increased illness, hospitalizations, and deaths within affected populations, particularly among vulnerable groups such as the elderly, immunocompromised individuals, and those with underlying health conditions.
- **Economic Impact:** Epidemics have significant economic consequences, including lost productivity, decreased consumer spending, disruptions to supply chains, and healthcare costs, affecting businesses, industries, and national economies.
- **Social Disruption:** Epidemics disrupt daily life, social activities, and community interactions, leading to fear, stigma, discrimination, and psychological distress among individuals and communities.
- **Long-Term Consequences:** Epidemics can have long-lasting effects on population health, healthcare systems, and societal resilience, influencing public health policies, emergency preparedness, and disease prevention efforts.

Risk Reduction

- **Disease Surveillance:** Establishing surveillance systems to monitor disease trends, detect outbreaks early, and track the spread of infectious diseases to inform public health interventions and control measures.
- **Vaccination and Immunization:** Implementing vaccination programs to prevent the spread of vaccine-preventable diseases and achieve herd immunity within populations, reducing the risk of epidemic outbreaks.
- **Infection Control Measures:** Promoting hand hygiene, respiratory etiquette, personal protective measures, and environmental sanitation to reduce disease transmission and mitigate epidemic risks.
- **Healthcare Preparedness:** Strengthening healthcare infrastructure, capacity, and response capabilities to effectively manage epidemics, provide medical care, and protect healthcare workers and patients.
- **Public Education and Communication:** Providing accurate, timely information, and communication to the public about epidemic risks, prevention strategies, and community mitigation measures to enhance awareness, cooperation, and compliance.
- **International Collaboration:** Enhancing international cooperation, coordination, and information sharing among countries, health organizations, and stakeholders to address global health threats, respond to epidemics, and strengthen pandemic preparedness and response efforts.

15. Loss of Biodiversity

Definition: Loss of biodiversity refers to the decline or extinction of plant and animal species, as well as the degradation or destruction of ecosystems and habitats. It encompasses the reduction in the variety and abundance of living organisms at genetic, species, and ecosystem levels.

Causes

- **Habitat Destruction:** Conversion of natural habitats for agriculture, urbanization, infrastructure development, and resource extraction leads to habitat loss and fragmentation, reducing available habitats for species.
- **Pollution:** Pollution from industrial activities, agricultural runoff, chemical contaminants, and plastic waste degrades air, water, and soil quality, negatively impacting biodiversity and ecosystem health.
- **Climate Change:** Rising temperatures, altered precipitation patterns, and extreme weather events disrupt ecosystems, shift habitats, and threaten species survival, contributing to biodiversity loss.
- **Overexploitation:** Unsustainable harvesting, hunting, fishing, logging, and wildlife trade exceed the capacity of ecosystems to regenerate, leading to declines in species populations and depletion of natural resources.

- **Invasive Species:** Introduction of non-native species disrupts native ecosystems, outcompetes native species for resources, and leads to species extinction, reducing biodiversity.
- **Land Use Change:** Changes in land use practices such as deforestation, agricultural expansion, and urban sprawl alter landscapes, fragment habitats, and degrade ecosystems, contributing to biodiversity loss.

Adverse Effects

- **Ecosystem Degradation:** Loss of biodiversity disrupts ecosystem functions, services, and resilience, affecting nutrient cycling, pollination, water purification, and climate regulation.
- **Species Extinction:** Biodiversity loss increases the risk of species extinction, leading to irreversible loss of genetic diversity, evolutionary potential, and ecological interactions.
- **Disruption of Ecosystem Services:** Decline in biodiversity reduces the capacity of ecosystems to provide essential services such as food production, disease regulation, carbon sequestration, and natural disaster mitigation.
- **Economic Impact:** Loss of biodiversity results in economic losses in sectors such as agriculture, forestry, fisheries, pharmaceuticals, and ecotourism, affecting livelihoods, industries, and national economies.
- **Social and Cultural Impacts:** Biodiversity loss threatens cultural heritage, traditional knowledge, and indigenous lifestyles dependent on natural resources, leading to social disruption, displacement, and loss of identity.

Risk Reduction

- **Habitat Conservation:** Protecting and restoring natural habitats, biodiversity hotspots, and critical ecosystems through conservation measures, protected areas, and restoration initiatives.
- **Sustainable Resource Management:** Implementing sustainable land use practices, ecosystem-based management, and conservation strategies to minimize the impacts of human activities on biodiversity and ecosystems.
- **Legal Protection:** Enacting laws, regulations, and international agreements to safeguard biodiversity, protect endangered species, and regulate activities that threaten ecosystems and habitats.
- **Public Awareness and Education:** Increasing awareness, appreciation, and understanding of biodiversity values, conservation issues, and sustainable lifestyles to promote stewardship and behaviour change.
- **Scientific Research and Monitoring:** Conducting research, monitoring programs, and biodiversity assessments to understand ecosystem dynamics, identify threats, and inform evidence-based conservation actions.

- **International Cooperation:** Enhancing international collaboration, cooperation, and funding for biodiversity conservation, research, and capacity-building initiatives to address global biodiversity challenges and achieve conservation goals.

Chapter-13

Disaster Management Strategies

Disaster Management efforts are geared towards disaster risk management.

Disaster Risk Management implies the systematic process of using administrative decisions, organisation, operational skills, and capacities to implement policies, strategies and coping capacities of the society and communities to lessen the impact of natural hazards and related environmental and technological disasters.

These comprise all forms all activities including structural and non- structural measures to avoid (prevention) or to limit (mitigation and preparedness) adverse effects of hazards.

There are three key stages of activities in disaster management:

Before a disaster: to reduce the potential for human, material, or environmental losses caused by hazards and to ensure that these losses are minimised when disaster strikes;

During a disaster: to ensure that the needs and provisions of victims are met to alleviate and minimise suffering;

After a disaster: to achieve rapid and durable recovery which does not reproduce the original vulnerable conditions.

Different Phases of Disaster Management

Phase	Description	Examples
Prevention	Actions taken to prevent disasters from occurring or to minimize their potential impact.	Enforcing building codes, implementing land-use planning regulations, conducting hazard assessments, promoting public awareness and education about disaster risks.
Mitigation	Measures aimed at reducing the severity of the potential impact of disasters by lessening the vulnerability of communities and infrastructure.	Constructing flood barriers, retrofitting buildings to withstand earthquakes, implementing drainage systems, establishing early warning systems.
Preparedness	Activities, plans, and resources put in place to effectively respond to disasters when they occur.	Developing emergency response plans, conducting training and drills for emergency responders and the public, stockpiling supplies, establishing communication networks, identifying evacuation routes.
Response	Immediate actions taken to provide assistance to affected individuals and communities during and immediately after a disaster.	Mobilizing emergency services, conducting search and rescue operations, providing medical care, distributing food and shelter, restoring essential services.
Recovery	Efforts to restore affected communities to a state of normalcy or better after the immediate crisis has passed.	Debris removal, rebuilding infrastructure, providing long-term housing solutions, offering psychological support to survivors, revitalizing the economy, implementing measures to reduce vulnerabilities.

Pre -Disaster Phase: Prevention, Mitigation & Preparedness

Prevention of Disaster

Prevention in disaster management involves measures taken to avoid or reduce the likelihood of a disaster occurring or its potential impacts. These measures aim to eliminate or minimize hazards, vulnerabilities, and risks within a community or environment. Prevention plays a critical role in reducing the need for emergency response and recovery efforts, ultimately saving lives, reducing property damage, and promoting sustainable development. There are various types of preventive measures implemented to achieve these objectives:

Types of Preventive Measures

(i) Structural Measures

- **Building Codes and Regulations:** Establishing and enforcing codes and regulations governing the design, construction, and maintenance of structures to withstand natural or man-made hazards, such as earthquakes, floods, hurricanes, and fires.
- **Infrastructure Development:** Constructing physical structures or systems, such as levees, dams, flood barriers, seawalls, stormwater management systems, and drainage channels, to mitigate the impact of hazards like floods, storm surges, and tsunamis.
- **Land-Use Planning:** Designating land use zones, regulating development in hazard-prone areas, and implementing setback requirements to reduce exposure to risks and prevent the construction of vulnerable structures in high-risk zones.

(ii) Non-Structural Measures

- **Environmental Conservation:** Preserving natural ecosystems, maintaining biodiversity, and protecting coastal wetlands, mangroves, forests, and natural buffers to mitigate the impact of hazards, such as floods, landslides, erosion, and storm surges.
- **Public Awareness and Education:** Conducting awareness campaigns, educational programs, workshops, and training sessions to inform communities about disaster risks, prevention measures, evacuation procedures, and emergency preparedness.
- **Early Warning Systems:** Establishing and operating systems to detect, monitor, and disseminate timely warnings and alerts about impending hazards, such as earthquakes, tsunamis, hurricanes, tornadoes, floods, and wildfires, to at-risk populations.

(iii) Technological Measures

- **Remote Sensing and Monitoring:** Using satellite imagery, aerial surveys, GIS (Geographic Information Systems), drones, and remote sensing technologies to assess and monitor changes in environmental conditions, identify potential hazards, and track disaster events in real-time.
- **Forecasting and Prediction:** Employing meteorological, hydrological, and geological data, models, and simulations to forecast the occurrence, intensity, and trajectory of hazardous events, enabling early preparation and response efforts.
- **Geotechnical Engineering Solutions:** Applying engineering techniques, such as soil stabilization, slope reinforcement, and landslide mitigation measures, to reduce the vulnerability of infrastructure and settlements to geological hazards, including landslides, soil liquefaction, and subsidence.

(iv) Legislative and Policy Measures

- **Disaster Risk Reduction (DRR) Policies:** Formulating and implementing policies, strategies, and action plans at national, regional, and local levels to integrate disaster risk reduction considerations into development planning, infrastructure investments, and decision-making processes.
- **Building Resilient Communities:** Promoting community-based approaches, participatory governance, and partnerships among governments, civil society organizations, academia, and the private sector to build resilient communities, enhance adaptive capacity, and foster social cohesion and inclusivity.

(v) International Cooperation and Collaboration

- **Global Initiatives and Frameworks:** Participating in international agreements, conventions, initiatives, and frameworks, such as the Sendai Framework for Disaster Risk Reduction, Paris Agreement on Climate Change, and Sustainable Development Goals (SDGs), to address transboundary risks, share knowledge, resources, and best practices, and enhance global cooperation in disaster prevention and mitigation efforts.

These types of preventive measures encompass a range of approaches and interventions aimed at reducing the risk of disasters and promoting the safety, well-being, and sustainability of communities and environments. Implementing a combination of structural, non-structural, technological, legislative, and international measures is essential for effective disaster prevention and risk reduction.

Disaster Preparedness

As disaster risk management focuses both on hazard and disaster, preparedness is also concerned to both the areas. Disaster management is a cyclic process which revolves around the four phases of loss reduction, i.e., mitigation, preparedness, response and recovery. Overall, preparedness aims that the different units of society like individual, household, institutions and communities cope with the disaster effectively and recover from the scourge of the disaster quickly. It includes measures to ensure life and property safety. The US National Emergency Management Association (NEMA) has identified twelve elements of preparedness. They are:

- Laws and Authorities
- Hazard identification and Risk Assessment
- Hazard mitigation
- Resource management
- Direction, Control and Coordination
- Communication and Warning
- Operations and Procedures
- Logistics and Facilities

- Training
- Exercise, Evaluation and Corrective Actions
- Crisis Communication, Education and Information
- Finance and Administration

The above stated twelve elements are part of the four phases of disaster risk management cycle

- US Federal Emergency Management Agency (FEMA) defines preparedness as– "The leadership, training, readiness and exercise support and technical and financial assistance to strengthen citizens, communities, state, local and tribal governments and professional emergency workers as they prepare for disasters, mitigate the effects of disaster, respond to community needs after the disaster and launch effective recovery efforts.

Community Based Disaster Preparedness

Community-based disaster preparedness (CBDP) is an approach that empowers local communities to proactively mitigate disaster impact through mobilizing resources, knowledge, and capacities. It focuses on assessing risks, developing response plans, and implementing measures to protect lives, livelihoods, and property. CBDP emphasizes local participation, collaboration, and ownership to strengthen community resilience against disasters.

Importance of Community Based Disaster Management (CBDM)

Disaster risk reduction measures achieve greater success when they directly engage the people most exposed to hazards. Community-based preparedness and early warning systems are crucial for saving lives, protecting property, and minimizing economic losses. Failure to comprehend the risk behaviour and cultural nuances of local communities can result in poorly designed preparedness measures, including ineffective early warning systems.

Active participation of local residents ensures self-reliance and ensures that emergency management plans are tailored to meet local needs. Indigenous knowledge and innovative approaches to hazard identification and mitigation also play pivotal roles. Conversely, disaster relief and recovery responses often fail to directly involve affected communities, leading to the provision of inappropriate and unsustainable forms of assistance.

Organized communities are inherently better equipped to demand downward accountability. A community-level focus aids in identifying vulnerable sections and tailoring interventions accordingly. Community-Based Disaster Management (CBDM) advocates for a bottom-up approach that complements top-down strategies, effectively addressing challenges and difficulties in disaster management.

Mitigation

Mitigation involves efforts to prevent hazards from turning into disasters or to minimize their effects if they occur. During this phase, actions such as establishing building codes, identifying hazardous areas, educating the public, and assessing vulnerability are commonly taken. The focus of mitigation is on implementing long-term measures to decrease or remove the risk of disasters. Mitigation activities fall into two categories:

Structural Mitigation: This involves making physical changes to infrastructure to reduce the likelihood or intensity of hazards. Examples include erosion control, sediment management, constructing earthquake-resistant and wind-resistant buildings, and building sandbag barriers to prevent flooding. These measures aim to safeguard life and property by understanding and addressing the expected stress from hazards.

Non-structural Mitigation: These are non-physical actions, such as policies and practices, aimed at increasing awareness of hazards. Measures include mental preparedness, training, insurance, planning, and discussions. Although indirect, these measures help to lessen the impact of disasters. Overall mitigation efforts, as identified by the Virtual University for Small States of the Commonwealth (VUSSC), include

- Hazard management
- Reducing vulnerability
- Economic diversification
- Political intervention
- Raising public awareness

During Disaster phase: Response and Recovery

Disaster Response:

Search and Rescue Operations

- Rapid deployment of search and rescue teams to locate and extract survivors from collapsed buildings, debris, or hazardous conditions.
- Utilization of specialized equipment such as search dogs, drones, and technical rescue teams.

Medical Assistance and Emergency Care

- Provision of immediate medical assistance, triage, and emergency care to injured individuals.
- Deployment of medical teams, field hospitals, and mobile clinics to treat casualties and prevent further health complications.

Shelter and Evacuation

- Establishment of temporary shelters, evacuation centers, or emergency housing for displaced individuals and families.
- Coordination of evacuation plans and transportation assistance for at-risk populations to safer locations.

Food, Water, and Essential supplies distribution

- Distribution of emergency relief supplies including food, water, blankets, hygiene kits, and basic necessities to affected communities.
- Coordination of supply chains and logistics to ensure timely delivery and equitable distribution of resources.

Communication and Information management

- Establishment of communication channels for disseminating emergency alerts, warnings, and updates to the public.
- Provision of accurate and accessible information on response efforts, safety guidelines, and available services.

Coordination and Collaboration

- Coordination among government agencies, emergency responders, NGOs, volunteers, and other stakeholders involved in the response.
- Collaboration with international organizations and neighbouring communities for resource mobilization and mutual aid.

Disaster Recovery

Damage Assessment and Needs Analysis

- Conducting comprehensive assessments to evaluate the extent of damage, loss, and needs of affected communities.
- Analysis of socio-economic impacts to prioritize recovery efforts and allocate resources effectively.

Infrastructure Repair and Reconstruction

- Repairing and rebuilding critical infrastructure such as roads, bridges, utilities, schools, hospitals, and communication networks.
- Incorporation of disaster-resilient design and construction practices to mitigate future risks.

Livelihood Support and Economic Recovery

- Implementation of livelihood support programs, vocational training, and income-generation projects to restore economic activities.
- Stimulus packages, microfinance initiatives, and small business grants to revive local economies and promote resilience.

Psychosocial support and Mental health services

- Provision of psychosocial support services, counselling, and trauma therapy to survivors, families, and responders.
- Creation of safe spaces, support groups, and community healing activities to address emotional and psychological needs.

Community Rehabilitation and Resilience Building

- Engaging communities in participatory planning and decision-making processes for recovery and rebuilding.
- Implementation of community-led projects for disaster risk reduction, environmental conservation, and sustainable development.

Resettlement and Housing Solutions

- Development of safe and resilient housing solutions, land tenure arrangements, and resettlement plans for displaced populations.
- Integration of housing infrastructure with social services, livelihood opportunities, and disaster preparedness measures.

Monitoring, Evaluation, and Learning

- Establishment of monitoring and evaluation mechanisms to track progress, assess impact, and ensure accountability in recovery efforts.
- Documentation of lessons learned, best practices, and innovations to inform future disaster response and resilience-building initiatives.

By addressing these components comprehensively and collaboratively, disaster response and recovery efforts can facilitate the restoration of affected communities, enhance resilience, and promote sustainable development in the aftermath of disasters.

Agencies involved in Disaster Management

National Disaster Management Authority of India (NDMA)

- It was established in 2005, under the Disaster Management Act 2005.
- The objective of NDMA is, to build a safer and disaster resilient India by a holistic, proactive, technology driven and sustainable development strategy.
- The NDMA is chaired by the Prime Minister of India and has a vice chairman with the status of Cabinet Minister and eight members with the status of Ministers of State.
- The NDMA Secretariat is headed by a secretary and deals with mitigation, preparedness, plans, reconstruction, community awareness and financial and administrative aspects.

National Disaster Management Plan (NDMP)

- It was released in 2016, it is the first ever national plan prepared in the country for disaster management.
- With National Disaster Management Plan (2016) India has aligned its National plan with the Sendai Framework for Disaster Risk Reduction 2015-2030, to which India is a signatory.
- The objective of the plan is to make India disaster resilient, achieve substantial disaster risk reduction. It aims to significantly decrease the losses of life, livelihoods, and assets in terms of economic, physical, social, cultural, and environmental. To maximize the ability to cope with disasters at all levels of administration as well as among communities.

State Disaster Management Authority (SDMA)

- At State level, State Disaster Management Authorities are established under Disaster Management Act 2005.
- SDMA is chaired by the Chief Minister of the State and has not more than eight members who are appointed by the Chief Minister.
- The SDMA prepares the state disaster management plan and implements the National Disaster Management Plan.

District Disaster Management Authority (DDMA)

- Under Disaster Management Act 2005, every State government shall establish a DDMA for every district in the State.
- The DDM Authority shall consist of:
- **Chairperson -** the Collector or District Magistrate or Deputy Commissioner act as Chairperson of DDMA.
- **Co-Chairperson -** is the elected representative of the local authority. In the Tribal Areas, the Chief Executive Member of the district council of autonomous district is the co-chairperson.
- There are not more than seven other members in DDMA.
- The Disaster Management Committee governed under District Magistrate will formulate village level disaster management plans for concern villages.
- The DDMA makes District Disaster Management Plan and implements the state Disaster Management Plan.

Local Authorities

Local authorities would include Panchayati Raj Institutions (PRI), Municipalities, District and Cantonment 11 Institutional and Legal Arrangements Boards, and Town Planning Authorities which control and manage civic services.

Post Disaster Phase: Reconstruction and Rehabilitation

Reconstruction

Reconstruction refers to the process of rebuilding and restoring infrastructure, communities, and livelihoods in the aftermath of a disaster. It involves physical, social, economic, and institutional measures aimed at returning affected areas to a state of normalcy or improved conditions.

Reconstruction is the comprehensive effort to rebuild and restore areas impacted by a disaster. It includes:

- **Physical Reconstruction:** Rebuilding damaged or destroyed infrastructure such as buildings, roads, bridges, schools, hospitals, and utilities. This involves employing engineering and construction techniques to ensure resilience against future disasters.
- **Social Reconstruction:** Restoring community structures, social networks, and support systems. This may involve constructing community facilities, restoring cultural landmarks, and promoting social cohesion and inclusion.
- **Economic Reconstruction:** Reviving economic activities, restoring livelihoods, and promoting sustainable development. This includes measures to support local businesses, create employment opportunities, and stimulate economic growth in affected areas.
- **Institutional Reconstruction:** Strengthening governance structures, legal frameworks, and institutional capacities to enhance disaster preparedness, response, and recovery. This involves reforming policies, regulations, and institutions to build resilience and promote good governance.
- **Environmental Reconstruction:** Rehabilitating natural ecosystems, mitigating environmental degradation, and promoting sustainable resource management. This includes reforestation, watershed management, and ecosystem restoration to improve resilience and reduce vulnerability to future disasters.

Reconstruction efforts require collaboration among government agencies, communities, NGOs, international organizations, and other stakeholders. It is essential to prioritize the needs of affected populations, promote inclusivity, and ensure sustainability in reconstruction projects. By addressing the multidimensional aspects of reconstruction, communities can rebuild stronger, more resilient, and sustainable futures in the aftermath of disasters

Rehabilitation

Rehabilitation involves restoring local services and enabling basic functions to resume after a disaster, facilitating the return to pre-disaster conditions. It encompasses various actions to repair physical damage, restore community facilities, revive economic activities, and support the psychological and social well-being of survivors. There are four major types of rehabilitation:

- **Physical Rehabilitation:** Involves rebuilding physical infrastructure such as houses, roads, water supply, and electricity. It also includes land acquisition for relocation, floodplain zoning, and environmental protection measures.
- **Social Rehabilitation:** Focuses on restoring social structures and supporting vulnerable groups like children, elderly, and marginalized communities. It includes constructing community centers, day care facilities, and old age homes to rebuild support systems.
- **Economic Rehabilitation:** Aims at restoring livelihoods and business continuity, emphasizing sustainable income generation. It involves measures to promote employment, income generation, and alternative livelihoods.
- **Psychological Rehabilitation:** Addresses the mental and emotional well-being of survivors, particularly those suffering from post-traumatic stress disorders. It includes stress management, trauma counselling, and psychological support services to help individuals cope with the psychological impacts of the disaster.

Early Warning and Disaster Communication

Early Warning Systems (EWS)

Components of EWS

- **Hazard Monitoring:** EWS involves continuous monitoring of various hazards such as weather events (storms, floods, hurricanes), geological events (earthquakes, tsunamis), biological threats (epidemics, pandemics), and technological hazards (industrial accidents, chemical spills).
- **Risk Assessment:** Once hazards are detected, risk assessment is conducted to evaluate their potential impact on communities. This includes analysing historical data, hazard maps, vulnerability assessments, and population exposure.
- **Warning Dissemination:** Timely and accurate warnings are disseminated to at-risk populations through multiple communication channels. These channels may include sirens, radio broadcasts, television alerts, mobile phone messages (SMS), social media platforms, community loudspeakers, and community networks.
- **Response Capacity:** EWS also involves building response capacity within communities and relevant authorities. This includes training emergency responders, conducting drills and exercises, establishing evacuation routes and shelters, and pre-positioning resources.

Types of Early Warning Systems

- **Hydro-meteorological Early Warning Systems:** These systems monitor weather and water-related hazards such as floods, storms, hurricanes, droughts, and heatwaves.
- **Geological Early Warning Systems:** These systems monitor geological hazards such as earthquakes, tsunamis, volcanic eruptions, landslides, and avalanches.
- **Biological Early Warning Systems:** These systems monitor biological threats such as epidemics, pandemics, disease outbreaks, and vector-borne diseases.
- **Technological Early Warning Systems:** These systems monitor technological hazards such as industrial accidents, chemical spills, nuclear incidents, and cyber-attacks.

Disaster Communication:

(i) Pre-Event Communication

- **Risk Awareness and Education:** Public education campaigns raise awareness about potential hazards, their impacts, and preparedness measures. These campaigns may include educational materials, workshops, seminars, and community outreach programs.
- **Community Engagement:** Engaging communities in disaster risk reduction activities fosters ownership, participation, and resilience. This may involve forming community-based disaster committees, conducting vulnerability assessments, and developing community action plans.

(ii) During-Event Communication

- **Warning Dissemination:** Rapid and effective dissemination of warnings to at-risk populations is crucial during disasters. This involves utilizing multiple communication channels and ensuring messages are clear, actionable, and accessible to diverse audiences.
- **Emergency Information Updates:** Providing real-time updates and instructions to the public during the event helps guide response actions. This may include evacuation orders, shelter locations, road closures, and safety precautions.

(iii) Post-Event Communication

- **Assessment and Reporting:** Conducting damage assessments and needs analyses to evaluate the impact of the disaster and identify priority areas for response and recovery.
- **Recovery Information:** Providing information on available assistance, support services, and recovery resources to affected individuals and

communities. This may include information on shelter, food distribution, medical services, and financial aid.

- **Lessons Learned and Preparedness Planning:** Analyzing the response effort to identify strengths, weaknesses, and lessons learned. This informs future preparedness planning and improvements to early warning systems and communication strategies.

Chapter-14

Sea Safety and Health

Sea safety and health are critical aspects of maritime operations and activities, encompassing measures and practices aimed at ensuring the well-being and safety of individuals involved in maritime endeavours, including crew members, passengers, and others working at sea.

Maritime Safety Regulations

- **International Standards**: Established by organizations like the International Maritime Organization (IMO) through conventions such as SOLAS, STCW, and MARPOL. These regulations set minimum standards for vessel construction, equipment, crew training, and environmental protection.
- **National Regulations**: Each country may have its own maritime regulatory agency responsible for implementing and enforcing regulations within its territorial waters. These regulations often align with international standards but may include additional requirements specific to national priorities and conditions.

Personal Protective Equipment (PPE)

- **Life Jackets**: Designed to keep individuals a float and provide thermal protection in cold water environments. They come in various types suitable for different activities and conditions.
- **Helmets**: Protect heads from impact injuries, particularly in situations where falling objects or collisions are possible, such as during heavy weather or cargo handling operations.

- **Gloves**: Essential for protecting hands from abrasions, cuts, and exposure to chemicals or harsh weather conditions. Different types of gloves are available depending on the specific hazards encountered.
- **Safety Harnesses**: Used to prevent falls overboard or from heights, particularly when working on decks or performing tasks aloft.

Emergency Preparedness and Response

- **Emergency Protocols**: Detailed procedures for responding to various emergencies, including fire, collision, grounding, medical emergencies, and man overboard situations.
- **Drills and Training**: Regular drills to ensure crew members are familiar with emergency procedures and can respond effectively under stressful conditions.
- **Emergency Equipment**: Availability and maintenance of emergency equipment such as fire extinguishers, life rafts, immersion suits, and first aid kits.

Health and Hygiene

- **Sanitation Standards**: Ensuring vessels have adequate facilities for personal hygiene, waste disposal, and clean water supply to prevent the spread of infectious diseases.
- **Food Safety**: Proper storage, handling, and preparation of food to prevent foodborne illnesses. Regular inspections and adherence to hygiene standards are essential.
- **Vector Control**: Measures to control pests such as rodents and insects, which can transmit diseases and compromise hygiene onboard.

Medical Facilities and Personnel

- **Medical Facilities**: Larger vessels should have dedicated medical facilities equipped to handle a range of medical emergencies. These may include a sick bay, medical equipment, and supplies.
- **Medical Personnel**: Trained medical professionals onboard, including doctors, nurses, or paramedics, capable of providing first aid, emergency care, and stabilization until further medical assistance can be obtained.

Mental Health Support

- **Counselling Services**: Access to confidential counselling services onboard or through telemedicine for addressing mental health issues such as depression, anxiety, and stress.
- **Stress Management Programs**: Providing resources and training to help individuals cope with the challenges of isolation, long working hours, and demanding conditions.

- **Peer Support Networks**: Establishing peer support networks or buddy systems to encourage open communication and mutual support among crew members.

Weather Monitoring and Navigation Safety

- **Weather Forecasting**: Utilizing advanced weather forecasting technologies to monitor weather conditions and anticipate potential hazards such as storms, high winds, and heavy seas.
- **Navigation Equipment**: Regular maintenance and calibration of navigation equipment such as GPS, radar, and electronic chart systems to ensure accurate positioning and safe navigation.

Safety Training and Education

- **Comprehensive Training Programs**: Providing comprehensive safety training covering a range of topics including emergency procedures, first aid, firefighting, survival techniques, and safety regulations.
- **Onboard Safety Induction**: Conducting safety inductions for all personnel joining a vessel to familiarize them with onboard procedures, emergency equipment, and safety policies.

Drug and Alcohol Policies

- **Zero Tolerance**: Enforcing strict policies prohibiting the use of drugs and alcohol onboard vessels to ensure the safety and well-being of crew members, passengers, and the integrity of operations.
- **Testing and Enforcement**: Conducting regular drug and alcohol testing to detect and deter substance abuse, with clear consequences for violations.

Environmental Protection

- **Pollution Prevention**: Implementing measures to minimize pollution from ship operations, including proper waste management, ballast water treatment, and reducing emissions.
- **Sustainable Practices**: Promoting sustainable practices such as fuel efficiency, alternative energy sources, and eco-friendly technologies to reduce the environmental impact of maritime activities.

Chapter-15

Acquaintance with Fire Fighting Devices

Fire Fighting

Fire on board is one of the major hazards in a vessel. A navigator should always be alert about the possibility of a fire. Prevention is better than fighting the fire. In the situation where there is fire accident the crew of the vessel must work together as a team to fight and survive. The navigator must know the type of fire and method of firefighting to select the right type of extinguisher and extinguishing agent. This chapter briefly gives an idea about the fire and fighting methods.

Chemistry of Fire

In the oxidation process a substance combines with oxygen and form new chemical. Burning is the rapid oxidation of millions of vapour molecules. The molecules oxidize by breaking apart into individual atoms and recombining with oxygen into new molecules. During the process of oxidation, heat energy is released to the environment. As the combustion is rapid oxidation the fuel combines with oxygen at a very high rate. Energy is released in the form of heat and light. This is called as fire.

Three things are essential to create a fire and continuous burning, they are,

- Fuel (some inflammable material to vaporize and burn)
- Heat (to raise the temperature of the fuel vapour to its ignition temperature)
- Air (Oxygen to combine with fuel vapour)

This is called a Fire Triangle. There will be fire, if this triangle is complete. Fire will not be formed, if any one side of this triangle is removed.

This technique is followed in firefighting. Fire could be extinguished by breaking the fire triangle. If fuel, oxygen, or heat is removed, the fire will be off.

- Remove the heat that cools down fire.
- Remove the oxygen that smooths the fire.
- Remove the fuel that starves the fire.

Types of Fires

Fire could be classified in to four types

- Class "A" fires (Alpha) involve common materials. Something made from wood, paper, cloth, rubber, and certain plastics. These objects produce an ash when burned.
- Class "B" fires (Bravo) involve flammable or combustible liquids, flammable gases, grease, pretty much anything that is a petroleum product.

Class of Fire	Type of Fire	Type of Extinguisher	Extinguisher Identification	Symbol
A	Ordinary combustibles: wood, paper, rubber, fabrics, and many plastics	Water, Dry Powder, Halon	A	
B	Flammable Liquids and Gases: gasoline, oils, paint, lacquer, and tar	Carbon Dioxide, Dry Powder Halon	B	
C	Fires involving Live Electrical Equipment	Carbon Dioxide, Dry Powder Halon	C	
D	Combustible Metals or Combustible Metal Alloys	Special Agents	D	No Picture Symbol D
K	Fires in Cooking Appliances that involve Combustible Cooking Media: Vegetable or Animal Oils and Fats		K	

- Class"C" fires (Charlie) involve electrical wire or instruments or equipment.
- Class"D" fires (Delta) involve combustible metals, such as Magnesium, Titanium, and Sodium (Mg, Ti, Na)
- Class "K" Fires involve cooking oils, fats, and grease typically found in commercial cooking appliances such as deep fryers. Class K fires are specific to kitchen fires and require extinguishers designed to handle high-temperature cooking oils.

Extinguishing Agents

An extinguishing agent is a substance that will remove any one side of fire triangle, causing the collapse of the triangle. Once the fire triangle is collapsed the fire will be under control and finally put off. Every extinguishing agent has specific action on fire, and one must be very careful while selecting the extinguishing agent for firefighting.

The action of extinguishing agent could be classified into,

Cooling - reduces the temperature of the fuel below its ignition temperature. This is a direct attack on the heat side of the fire triangle

Smothering -separates the fuel from the oxygen. This can be considered as an attack on the edge of the fire triangle where the fuel and oxygen sides meet.

Oxygen dilution-reduces the amount of available oxygen below that needed to sustain combustion. This is an attack on the oxygen side of the triangle

Fire Extinguishing Agents

The following are the common fire extinguishing agents

(i) Liquids

- **Water:** Water will act as cooling agent that absorbs the heat and cools the burning materials. Water is the commonly available extinguishing agents.
- **Foam:** Foam is a mass of bubbles that cover the fire like blanket and extinguishes the fire by smothering. Mixing water and a foam-making agent (foam concentrate) produces bubbles Chemical foam could be produced by mixing an alkali (usually Sodium bicarbonate) with an acid (usually Aluminium sulphate) in water.

When chemical foam was first introduced, these substances were stored in separate containers. These two chemicals are packed separately in a sealed, airtight container. When these chemicals are intentionally mixed, they react and foam or froth of bubbles filled with Carbon dioxide gas are produced.

(ii) Gases

Carbon Dioxide (CO_2): Carbon dioxide extinguishes fire by smothering. It dilutes the air surrounding the fire until the oxygen content is too low to support combustion. For this reason, it is effective on Class B fires, where the main consideration is to keep the flammable vapours separated from oxygen in the air.

(iii) Solid (Dry Chemical)

Sodium bicarbonate is commonly used dry chemical extinguishing agent. It is widely used because it is the most economical extinguishing agent. It is particularly effective on animal fats and vegetable oils because it chemically changes these substances into non-flammable soaps.

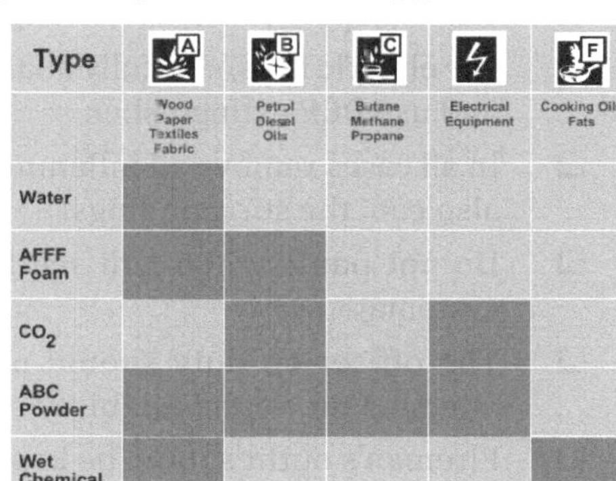

Fire Fighting Appliances

Various fire extinguishers exist for use on different fire sources. When dealing with a fire which is being fuelled by burning wood or paper, you may act very differently from the way you would on an electrical fire. Obviously, water cannot be used on an electrical fire as water conducts electricity and could put you in grave danger if mixed with live electrics. This means you must choose a suitable fire extinguisher for the job in hand – even in the most perilous of emergency situations.

Every fire extinguisher has a colour-coded band that gives a quick reference as to what is in the container. It is essential that you learn these codes and the situations in which they are suitable to be used.

The principal precautions necessary to prevent fires on board fishing vessels:

- Awareness and training for fishermen about basics of fire and fire prevention.
- Keeping the fire extinguishers in suitable location.
- Maintenance of engine room and oil system free from leaks.
- Periodical checking of the bilge for oil and grease leaks.
- Clean the spilled fuel or oil immediately and properly dispose the wastes.
- Be alert for suspicious odours and fumes.
- Check for abrasions, cracked wiring, or pinholes in fuel and lube oil lines

Fire on Board

When there is fire onboard:

- Shout FIRE, FIRE, FIRE.
- Give the fire alarm.
- Find out what type of fire it is.
- Alert the persons in the rest cabin and other places.
- Close all the ventilators and blowers to restrict the supply of air
- Start fighting with the fire, the movement it is detected.
- Use the appropriate fire extinguisher
- For electric fires, carefully analyse the situation, cut off the power supply and use DCP extinguisher.
- In all cases remove all inflammable materials from the surroundings and also cool the surroundings.
- Do not panic when you see fire, act wisely and do every action in a systematic way.
- The officer on duty should monitor every action in the ship without affecting the vessels stability.
- Fireman's outfit should be kept in all the vessels.

Fireman's Outfits

- ❑ Fireproof asbestos suit.
- ❑ Boots and gloves
- ❑ A rigid helmet.
- ❑ An electric safely lamp.
- ❑ An axe
- ❑ A smoke helmet or smoke mask

Fire Blankets

- ❑ Used for smothering small fires or wrapping around individuals during a fire emergency.
- ❑ Know how to deploy a fire blanket swiftly and effectively.

Fire Hose Reels

- ❑ Identify fire hose reel locations and understand their operation.
- ❑ Learn how to turn on/off and control water flow from a fire hose reel.
- ❑ Be aware of safety precautions, including nozzle kickback, when using hose reels.

Fire Alarms and Detection Systems

- ❑ Recognize manual call points, smoke detectors, heat detectors, and flame detectors.
- ❑ Understand evacuation procedures upon activation of fire alarms.

Fire Sprinkler Systems

- ❑ These automatic fire suppression systems are installed in buildings and activate when they detect heat from a fire.
- ❑ They release water or other extinguishing agents to control or extinguish the fire.

Smoke Detectors and Fire Alarms

- ❑ While not directly used to fight fires, these devices play a crucial role in fire safety by detecting smoke or fire and alerting occupants to evacuate the building.

Fire Hydrants

- ❑ Connected to the municipal water supply, fire hydrants provide a ready source of water for firefighters to use in extinguishing fires.

Firefighting Gear and Equipment

- ❑ This includes protective clothing (such as firefighting suits, helmets, gloves, and boots), breathing apparatus (like SCBA - Self-Contained Breathing

Apparatus), and various tools used by firefighters to perform their duties safely and effectively.

Emergency Escape Routes

- Clearly mark all emergency escape routes with illuminated signs to ensure visibility in low-light conditions.
- Regularly inspect and maintain emergency exits to ensure they are clear of obstructions and operational at all times.

Personal Protective Equipment (PPE)

- Understand the importance of wearing appropriate PPE during fire emergencies.
- Regularly inspect and maintain PPE to ensure effectiveness.

Fire Safety Regulations

- Adhere to relevant fire safety regulations and guidelines.
- Seek fire safety training to stay informed about regulations and best practices.

Emergency Response Procedures

- Understand the roles and responsibilities during a fire emergency, encompassing communication and assistance.
- Employ clear and effective communication techniques to coordinate responses efficiently.

Chapter-16

Lifesaving Appliances and First-Aid

A. Life Saving Application

(i) Life Jackets

- Types include inherently buoyant (foam-filled) and inflatable life jackets.
- Should be appropriately sized and fitted for each individual.
- Often equipped with reflective strips and whistles for visibility and signalling.

(ii) Life Rafts

- Constructed of durable, buoyant materials with canopies for protection against elements.
- Equipped with survival supplies such as water, food rations, first aid kits, and signalling devices.
- Capacity varies based on the size and intended use of the life raft.

(iii) Lifebuoys

- Typically made of buoyant materials like foam or plastic.
- Attached to a rope for throwing to individuals in distress.
- Some models may include lights or reflective tape for visibility in low light conditions.

(iv) Rescue Boats

- Rigid-hulled or inflatable vessels designed for rapid deployment in emergency situations.
- Equipped with propulsion systems (engine or oars), navigation lights, and communication devices.
- Trained personnel operate rescue boats to perform rescues safely and efficiently.

(v) Fire Extinguishers

- Classifications include water, foam, CO_2, dry chemical, and wet chemical extinguishers, each suitable for different types of fires.
- Regular inspection, maintenance, and training on proper usage are essential.
- Location and accessibility are key factors in effective fire prevention and response.

(vi) Emergency Oxygen

- Portable oxygen tanks with regulators and masks for delivering oxygen to individuals experiencing respiratory distress.
- Can be crucial in situations such as near drowning, suffocation, or heart attacks to support vital functions until advanced medical care is available.

(vii) Defibrillators (AED)

- Automated devices that analyse heart rhythms and deliver electrical shocks to restore normal cardiac function in cases of sudden cardiac arrest.
- Designed for ease of use by lay persons with audio and visual prompts guiding the user through the process.
- Rapid defibrillation within minutes of cardiac arrest significantly improves survival rates.

(viii) Emergency Evacuation Systems

- Aircraft evacuation slides are inflatable slides deployed from aircraft doors in emergencies, facilitating rapid evacuation.
- Building evacuation chutes provide safe descent from elevated locations during fires or other emergencies.
- Regular drills and familiarization with evacuation procedures are critical for efficient evacuation.

B. First Aid

(i)CPR (Cardiopulmonary Resuscitation)

- Involves cycles of chest compressions and rescue breaths to maintain circulation and oxygenation.

- The compression-to-ventilation ratio and technique may vary based on the rescuer's training level and the victim's condition.

(ii) Bandages and Dressings

- Various types include adhesive bandages, gauze pads, roller bandages, and triangular bandages.
- Used to control bleeding, protect wounds, and promote healing.
- Proper wound cleaning and dressing techniques are essential to prevent infection.

(iii) Splints

- Devices used to immobilize fractured or injured limbs to prevent further damage and reduce pain.
- Can be improvised from materials like wooden boards, cardboard, or even rolled-up magazines in emergencies.

(iv) Burn Dressings

- Cool, moist dressings or gels applied to burns to soothe pain, reduce inflammation, and prevent infection.
- Should be applied promptly to minimize tissue damage and alleviate discomfort.

(v) Antiseptics and Disinfectants

- Common examples include alcohol swabs, hydrogen peroxide, and iodine solutions.
- Used to clean wounds and prevent infection by killing or inhibiting the growth of microorganisms.

(vi) Epinephrine Auto-injectors

- Pre-filled devices containing epinephrine (adrenaline) for immediate administration in cases of severe allergic reactions (anaphylaxis).
- Administered intramuscularly into the thigh and can rapidly reverse symptoms such as difficulty breathing and low blood pressure.

(vii) Tourniquets

- Applied proximal to a severe limb injury to control bleeding when direct pressure or other methods are ineffective.
- Should be applied with caution to avoid excessive tissue damage and released as soon as professional medical help is available.

(viii) Airway Management Devices

- Nasopharyngeal airways are soft, flexible tubes inserted through the nostrils to maintain an open airway in unconscious or injured individuals.

- Pocket masks with one-way valves allow rescuers to provide rescue breaths while minimizing the risk of cross-contamination.

(ix) Emergency Response Training

- Includes courses such as Basic Life Support (BLS), Advanced Cardiovascular Life Support (ACLS), and First Aid/CPR/AED training.
- Hands-on practice and scenario-based training enhance skills retention and preparedness for real-life emergencies.
- Regular refresher courses and updates ensure that responders remain proficient in their skills and knowledge.

C. Use of Distress Signals and Technologies

Distress signals and technologies play a crucial role in alerting authorities and requesting assistance during emergencies, particularly in remote or hazardous environments. Here are some of the primary uses of distress signals and technologies:

(a) Maritime Distress Signals

i. **EPIRBs (Emergency Position Indicating Radio Beacons):** Automatically transmit distress signals via satellite to search and rescue authorities, indicating the vessel's precise location in case of maritime emergencies such as sinking, fire, or abandonment.

ii. **Flares:** Pyrotechnic devices used to visually signal distress to nearby vessels or aircraft. They come in various types, including handheld, parachute, and rocket flares, and are essential for attracting attention in low-visibility conditions or at night.

iii. **DSC (Digital Selective Calling):** Integrated into marine radios, DSC allows vessels to send distress alerts with precise GPS coordinates to nearby vessels and maritime rescue coordination centers, facilitating rapid response and assistance.

(b) Aviation Distress Signals

i. **Emergency Locator Transmitters (ELTs):** Automatically activate upon impact or manually by aircraft crew in the event of a crash, transmitting distress signals on designated frequencies to aid search and rescue efforts.

ii. **Emergency Position-Indicating Radio Beacons (EPIRBs):** Similar to maritime EPIRBs, aviation EPIRBs are designed for aircraft and can transmit distress signals via satellite to alert search and rescue authorities of an emergency.

iii. **Visual Signals:** Aircraft distress signals may include visual cues such as flashing lights, coloured smoke, or distress flags deployed from the aircraft to attract attention from ground or air search teams.

(c) Land-based Distress Signals

iv. **Cell Phones:** Widely used to call emergency services and report distress situations on land. Modern smartphones often have GPS capabilities, allowing authorities to track the caller's location accurately.

v. **Personal Locator Beacons (PLBs):** Portable distress beacons carried by individuals during outdoor activities such as hiking, mountaineering, or boating. PLBs transmit distress signals via satellite to request assistance in remote areas where cell phone coverage is unavailable.

vi. **Flares and Signal Mirrors:** Used by individuals stranded or lost in wilderness or remote areas to signal distress to potential rescuers. Flares produce bright light, while signal mirrors reflect sunlight to attract attention from search aircraft or ground teams.

(d) Personal Distress Beacons

i. **PLBs (Personal Locator Beacons):** Compact, portable devices that individuals can carry to transmit distress signals via satellite in emergency situations, providing their precise location for search and rescue teams.

ii. **Satellite Messengers:** Devices equipped with GPS tracking and satellite communication capabilities, allowing users to send distress messages, share their location, and communicate with emergency contacts or rescue services from remote areas beyond cellular coverage.

D. Relief and Rehabilitation Measures

(i) Emergency Shelter and Aid Distribution

- ❑ Provide immediate shelter, food, water, and essential supplies to displaced populations following disasters or conflicts.
- ❑ Coordinate with humanitarian organizations and government agencies to ensure rapid response and distribution of aid to affected communities.

(ii) Medical Assistance and Healthcare Services

- ❑ Deploy medical teams and establish temporary clinics or field hospitals to provide emergency medical care to survivors.
- ❑ Conduct vaccination campaigns, disease surveillance, and hygiene promotion activities to prevent the spread of infectious diseases in crowded or unsanitary conditions.

(iii) Psychosocial Support and Mental Health Services

- ❑ Offer counseling, support groups, and psychotherapy to individuals affected by trauma, grief, or loss.
- ❑ Train local community members as lay counselors to provide peer support and mental health education to vulnerable populations.

(iv) Education and Child Protection Programs

- Reopen schools and establish safe learning spaces for children affected by disasters or conflicts, prioritizing psychosocial support and trauma-informed teaching practices.
- Implement child protection measures to prevent exploitation, abuse, and separation of children from their families during emergencies.

(v) Livelihood Support and Economic Recovery

- Provide cash assistance, livelihood training, and small business grants to help affected communities rebuild their livelihoods and regain economic stability.
- Support agricultural recovery, vocational skills development, and employment opportunities to enhance resilience and promote sustainable recovery.

E. Trauma Counselling

(i) Initial Assessment and Psychological First Aid

- Conduct rapid assessments to identify individuals at risk of psychological distress and provide immediate psychological first aid.
- Offer compassionate listening, reassurance, and practical support to survivors to address their immediate emotional and practical needs.

(ii) Trauma-Informed Counselling and Therapy

- Provide evidence-based trauma-focused therapies, such as cognitive-behavioral therapy (CBT) and eye movement desensitization and reprocessing (EMDR), to address trauma symptoms and promote healing.
- Utilize culturally sensitive approaches and respect survivors' autonomy and coping mechanisms in the therapeutic process.

(iii) Group Support and Peer Counselling

- Facilitate support groups and peer-led counseling sessions to foster social connections, mutual understanding, and solidarity among survivors.
- Encourage sharing of experiences, coping strategies, and resilience-building activities to promote emotional healing and community cohesion.

(iv) Community-Based Mental Health Promotion

- Train community leaders, teachers, and health workers in basic counseling skills, trauma awareness, and psychosocial support techniques.
- Conduct community outreach activities, awareness campaigns, and stigma reduction initiatives to promote mental health literacy and resilience at the grassroots level.

Glossary

Aerosols: Tiny particles suspended in the atmosphere, including dust, pollutants, and volcanic ash, which can influence climate by scattering or absorbing sunlight and affecting cloud formation.

Albedo: The reflectivity of a surface, indicating the proportion of incoming solar radiation that is reflected back into space rather than absorbed.

Atmosphere: The layer of gases surrounding a planet, particularly Earth, extending from its surface into space.

Blizzard: A severe winter storm characterized by strong winds, low temperatures, and blowing snow, reducing visibility and creating hazardous conditions.

Climate Change: Long-term shifts in temperature, precipitation patterns, sea levels, and other climate variables, primarily driven by human activities such as burning fossil fuels and deforestation.

Climate Modelling: The use of mathematical models and computer simulations to predict future climate conditions based on past observations, physical principles, and greenhouse gas emissions scenarios.

Climate Variability: Natural fluctuations in climate patterns occurring over various timescales, ranging from years to centuries, influenced by factors like solar activity, volcanic eruptions, and ocean currents.

Climate: The long-term patterns and averages of weather conditions in a particular region over an extended period, typically 30 years or more.

Cryosphere: The frozen components of Earth's surface, including glaciers, ice caps, ice sheets, sea ice, and permafrost, crucial for regulating climate and sea level.

Dendrochronology: The study of tree rings to determine past climate conditions, including temperature, precipitation, and wildfire frequency.

Disaster: A sudden, catastrophic event causing widespread destruction, disruption, and loss of life, property, and livelihoods.

Drought: A prolonged period of abnormally low precipitation, leading to water shortages, crop damage, and environmental impacts.

Early Warning System: A system of monitoring, detection, and communication designed to provide advance notice of impending disasters, allowing time for preparation and evacuation.

El Niño: A climate phenomenon characterized by warmer-than-average sea surface temperatures in the central and eastern Pacific Ocean, often leading to weather disruptions worldwide.

Emergency Preparedness: The process of planning, organizing, and training to respond effectively to emergencies, minimizing their impact and protecting lives and property.

Flood: An overflow of water onto land that is usually dry, caused by heavy rainfall, snowmelt, or the overflow of rivers and lakes.

Front: The boundary between two air masses with different temperatures or moisture levels, often associated with changes in weather conditions.

Frontal lifting: It is a weather phenomenon that occurs when two different air masses with different temperatures, densities and humidity levels meet and one air mass is forced to rise over the other.

Glacier: A large mass of ice formed from compacted snow, moving slowly downhill under the influence of gravity, sensitive to changes in temperature and precipitation.

Global Warming: The gradual increase in Earth's average temperature due to the enhanced greenhouse effect caused by human activities.

Greenhouse Effect: The process by which greenhouse gases in the Earth's atmosphere trap heat from the Sun, warming the planet's surface and making it habitable.

Heatwave: A prolonged period of excessively hot weather, often accompanied by high humidity, posing health risks to humans and animals.

Humanitarian Aid: Assistance provided to people affected by disasters or crises, including food, water, shelter, medical care, and other essential services, often delivered by humanitarian organizations and relief agencies.

Humidity: The amount of water vapor present in the atmosphere, expressed as a percentage of the maximum amount the air could hold at a given temperature.

Hurricane: A tropical cyclone with sustained winds of 74 miles per hour (119 kilometers per hour) or more, typically forming over the Atlantic Ocean or eastern Pacific Ocean.

Hydrological Cycle: The continuous movement of water on, above, and below the Earth's surface, driven by solar energy, evaporation, condensation, precipitation, and runoff.

Ice Age: A period of geological time characterized by extensive ice sheets covering large portions of Earth's surface, typically lasting millions of years.

Ice Core: Cylindrical samples of ice drilled from glaciers and ice sheets, containing layers of snow and air bubbles that provide valuable information about past climate conditions.

Incident Command System (ICS): A standardized management structure used by emergency responders to coordinate and manage on-site operations during complex incidents or disasters.

La Niña: The opposite of El Niño, characterized by cooler-than-average sea surface temperatures in the central and eastern Pacific Ocean, influencing global weather patterns.

Microclimate: The climate conditions prevailing in a small, localized area, such as a garden, park, or urban area, influenced by factors like topography, vegetation, and human activities.

Mitigation: Actions taken to reduce or prevent the severity, impact, and recurrence of disasters, including measures such as land-use planning, infrastructure improvements, and hazard-resistant construction.

Monsoon: A seasonal wind pattern characterized by heavy rainfall and distinct wet and dry seasons, commonly observed in South Asia, Southeast Asia, and parts of Africa.

Ozone Layer: A region of the Earth's stratosphere containing a high concentration of ozone O_3 that absorbs and blocks ultraviolet (UV) radiation from the Sun, protecting life on Earth from harmful UV rays.

Orographic lifting: when a moving air mass meet a geographical feature such as a mountain, some of the air will have no option but to rise up over it.

Paleoclimate: The climate conditions prevailing on Earth in the geological past, reconstructed using evidence from ice cores, sediment layers, fossil records, and other sources.

Permafrost: Ground that remains permanently frozen, typically found in polar regions and high mountainous areas, vulnerable to thawing due to climate change.

Polar Amplification: The phenomenon whereby temperature changes are amplified in polar regions compared to global averages, leading to faster rates of warming and melting of ice.

Precipitation: Any form of water, such as rain, snow, sleet, or hail, that falls from the atmosphere to the Earth's surface.

Pressure: The force exerted by the weight of the atmosphere above a given point, typically measured in millibars or inches of mercury.

Recovery: The process of rebuilding, restoring, and rehabilitating affected communities and infrastructure following a disaster, aimed at returning to normalcy and reducing vulnerability to future disasters.

Resilience: The ability of individuals, communities, and systems to withstand, adapt to, and recover from disasters, maintaining essential functions and minimizing long-term impacts.

Response: The coordinated efforts to provide immediate assistance and relief to affected populations during and immediately after a disaster, including search and rescue, medical care, and shelter provision.

Risk Assessment: The systematic evaluation of potential hazards, vulnerabilities, and exposures to determine the likelihood and potential consequences of disasters.

Temperature: A measure of the average kinetic energy of particles in a substance or environment, indicating how hot or cold it is.

Thermohaline Circulation: The global ocean circulation driven by differences in temperature and salinity, influencing climate patterns and redistributing heat around the world.

Thunderstorm: A localized weather phenomenon characterized by lightning, thunder, heavy rain, and sometimes hail or strong winds.

Tornado: A violently rotating column of air extending from a thunderstorm to the ground, capable of causing significant damage.

Tropical Cyclone: A large, rotating storm system with low pressure and organized convection, typically forming over warm ocean waters and characterized by strong winds and heavy rainfall.

Urban Heat Island: An urban area or metropolitan area significantly warmer than its surrounding rural areas due to human activities, such as building materials, vehicle emissions, and heat generation.

Weather: The state of the atmosphere at a specific time and place, including temperature, humidity, precipitation, wind speed, and atmospheric pressure.

Wind: The horizontal movement of air across the Earth's surface, caused by differences in air pressure.

References

Barry, R. G., & Chorley, R. J. (2018). Atmosphere, Weather and Climate. Routledge.

Gaillard, J. C. (2020). Climate Change and Disaster Risk Management. Routledge.

Henson, R. (2020). The Thinking Person's Guide to Climate Change. American Meteorological Society.

Kelman, I., Gaillard, J. C., & Lewis, J. (2020). Handbook of Hazards and Disaster Risk Reduction. Routledge.

Kundzewicz, Z. W., & Krysanova, V. (Eds.). (2019). Climate Change, Hazards and Adaptation Options: Progress and Challenges in South-Eastern Europe. Springer.

Naylor, S. (2018). Introduction to Disaster Management. Routledge.

Stull, R. (2015). Practical Meteorology: An Algebra-based Survey of Atmospheric Science. University Science Books.